D1568008

GAUGE FIELDS

INTRODUCTION TO QUANTUM THEORY

FRONTIERS IN PHYSICS: A Lecture Note and Reprint Series

David Pines, Editor

Volumes of the Series published from 1961 to 1973 are not officially numbered. The parenthetical numbers shown are designed to aid librarians and bibliographers to check the completeness of their holdings.

(1)	N. Bloembergen	Nuclear Magnetic Relaxation: A Reprint Volume, 1961
(2)	G. F. Chew	*S*-Matrix Theory of Strong Interactions: A Lecture Note and Reprint Volume, 1961
(3)	R. P. Feynman	Quantum Electrodynamics: A Lecture Note and Reprint Volume, 1961 (6th printing, 1980)
(4)	R. P. Feynman	The Theory of Fundamental Processes: A Lecture Note Volume, 1961 (5th printing, 1979)
(5)	L. Van Hove, N. M. Hugenholtz, and L. P. Howland	Problems in Quantum Theory of Many-Particle Systems: A Lecture Note and Reprint Volume, 1961
(6)	D. Pines	The Many-Body Problem: A Lecture Note and Reprint Volume, 1961 (5th printing, 1979)
(7)	H. Frauenfelder	The Mössbauer Effect: A Review—with a Collection of Reprints, 1962
(8)	L. P. Kadanoff and G. Baym	Quantum Statistical Mechanics: Green's Function Methods in Equilibrium and Nonequilibrium Problems, 1962 (3rd printing, 1976)
(9)	G. E. Pake	Paramagnetic Resonance: An Introductory Monograph, 1962 [cr. (42)—2nd edition]
(10)	P. W. Anderson	Concepts in Solids: Lectures on the Theory of Solids, 1963 (4th printing, 1978)
(11)	S. C. Frautschi	Regge Poles and *S*-Matrix Theory, 1963
(12)	R. Hofstadter	Electron Scattering and Nuclear and Nucleon Structure: A Collection of Reprints with an Introduction, 1963
(13)	A. M. Lane	Nuclear Theory: Pairing Force Correlations to Collective Motion, 1964
(14)	R. Omnès and M. Froissart	Mandelstam Theory and Regge Poles: An Introduction for Experimentalists, 1963
(15)	E. J. Squires	Complex Angular Momenta and Particle Physics: A Lecture Note and Reprint Volume, 1963
(16)	H. L. Frisch and J. L. Lebowitz	The Equilibrium Theory of Classical Fluids: A Lecture Note and Reprint Volume, 1964
(17)	M. Gell-Mann and Y. Ne'eman	The Eightfold Way: (A Review—with a Collection of Reprints), 1964
(18)	M. Jacob and G. F. Chew	Strong-Interaction Physics: A Lecture Note Volume, 1964
(19)	P. Nozières	Theory of Interacting Fermi Systems, 1964

FRONTIERS IN PHYSICS: A Lecture Note and Reprint Series

David Pines, Editor *(continued)*

(20)	J. R. Schrieffer	Theory of Superconductivity, 1964 (2nd printing, 1971)
(21)	N. Bloembergen	Nonlinear Optics: A Lecture Note and Reprint Volume, 1965 (3rd printing, with addenda and corrections, 1977)
(22)	R. Brout	Phase Transitions, 1965
(23)	I. M. Khalatnikov	An Introduction to the Theory of Superfluidity, 1965
(24)	P. G. de Gennes	Superconductivity of Metals and Alloys, 1966
(25)	W. A. Harrison	Pseudopotentials in the Theory of Metals, 1966 (2nd printing, 1971)
(26)	V. Barger and D. Cline	Phenomenological Theories of High Energy Scattering: An Experimental Evaluation, 1967
(27)	P. Choquard	The Anharmonic Crystal, 1967
(28)	T. Loucks	Augmented Plane Wave Method: A Guide to Performing Electronic Structure Calculations—A Lecture Note and Reprint Volume, 1967
(29)	T. Ne'eman	Algebraic Theory of Particle Physics: Hadron Dynamics in Terms of Unitary Spin Currents, 1967
(30)	S. L. Adler and R. F. Dashen	Current Algebras and Applications to Particle Physics, 1968
(31)	A. B. Migdal	Nuclear Theory: The Quasiparticle Method, 1968
(32)	J. J. J. Kokkedee	The Quark Model, 1969
(33)	A. B. Migdal and V. Krainov	Approximation Methods in Quantum Mechanics, 1969
(34)	R. Z. Sagdeev and A. A. Galeev	Nonlinear Plasma Theory, 1969
(35)	J. Schwinger	Quantum Kinematics and Dynamics, 1970
(36)	R. P. Feynman	Statistical Mechanics: A Set of Lectures, 1972 (6th printing, 1981)
(37)	R. P. Feynman	Photo-Hadron Interactions, 1972
(38)	E. R. Caianiello	Combinatorics and Renormalization in Quantum Field Theory, 1973
(39)	G. B. Field, H. Arp, and J. N. Bahcall	The Redshift Controversy, 1973 (2nd printing, 1976)
(40)	D. Horn and F. Zachariasen	Hadron Physics at Very High Energies, 1973
(41)	S. Ichimaru	Basic Principles of Plasma Physics: A Statistical Approach, 1973 (2nd printing, with revisions, 1980)
(42)	G. E. Pake and T. L. Estle	The Physical Principles of Electron Paramagnetic Resonance, 2nd Edition, completely revised, enlarged, and reset, 1973 [cf. (9)—1st edition]

FRONTIERS IN PHYSICS: A Lecture Note and Reprint Series

David Pines, Editor *(continued)*

Volumes published from 1974 onward are being numbered as an integral part of the bibliography:

Number		
43	R. C. Davidson	Theory of Nonneutral Plasmas, 1974
44	S. Doniach and E. H. Sondheimer	Green's Functions for Solid State Physicists, 1974
45	P. H. Frampton	Dual Resonance Models, 1974
46	S. K. Ma	Modern Theory of Critical Phenomena, 1976
47	D. Forster	Hydrodynamic Fluctuations, Broken Symmetry, and Correlation Functions, 1975
48	A. B. Migdal	Qualitative Methods in Quantum Theory, 1977
49	S. W. Lovesey	Condensed Matter Physics: Dynamic Correlations, 1980
50	L. D. Faddeev and A. A. Slavnov	Gauge Fields: Introduction to Quantum Theory, 1980

Other Numbers in preparation.

GAUGE FIELDS

INTRODUCTION TO QUANTUM THEORY

L. D. FADDEEV *Academy of Sciences U.S.S.R.*
Steclov Mathematical Institute, Leningrad

A. A. SLAVNOV *Academy of Sciences U.S.S.R.*
Steclov Mathematical Institute, Moscow

Translated from the Russian Edition by
D. B. PONTECORVO
Joint Institute for Nuclear Research, Dubna

1980
THE BENJAMIN/CUMMINGS PUBLISHING COMPANY, INC.
ADVANCED BOOK PROGRAM
Reading, Massachusetts

London • Amsterdam • Don Mills, Ontario • Sydney • Tokyo

WILLIAM MADISON RANDALL LIBRARY UNC AT WILMINGTON

CODEN: FRHPA

L. D. Faddeev and A. A. Slavnov

Gauge Fields: Introduction to Quantum Theory

Originally published in 1978 as ВВЕДЕНИЕ В КВАНТОВУЮ ТЕОРИЮ КАЛИБРОВОЧНЫХ ПОЛЕЙ by Izdatel'stvo "Nauka," Moscow

1st printing, 1980
2nd printing, 1982

ISBN 0-8053-9016-2
Library of Congress Catalog Card Number 80-14006

Copyright © 1980 The Benjamin/Cummings Publishing Company, Inc.
Published simultaneously in Canada.

All rights reserved. No part of this publication may be reproduced, stored in a retrieval system, or transmitted, in any form or by any means, electronic, mechanical, photocopying, recording, or otherwise, without the prior written permission of the publisher, The Benjamin/Cummings Publishing Company, Inc., Advanced Book Program, Reading, Massachusetts 01867, U.S.A.

Printed in the United States of America

BCDEFGHIJ-AL-898765432

QC793
.3
.F5
.S5213

CONTENTS

224523

EDITOR'S FOREWORD

The problem of communicating in a coherent fashion recent developments in the most exciting and active fields of physics seems particularly pressing today. The enormous growth in the number of physicists has tended to make the familiar channels of communication considerably less effective. It has become increasingly difficult for experts in a given field to keep up with the current literature; the novice can only be confused. What is needed is both a consistent account of a field and the presentation of a definite "point of view" concerning it. Formal monographs cannot meet such a need in a rapidly developing field, and, perhaps more important, the review article seems to have fallen into disfavor. Indeed, it would seem that the people most actively engaged in developing a given field are the people least likely to write at length about it.

FRONTIERS IN PHYSICS has been conceived in an effort to improve the situation in several ways. Leading physicists today frequently give a series of lectures, a graduate seminar, or a graduate course in their special fields of interest. Such lectures serve to summarize the present status of a rapidly developing field and may well constitute the only coherent account available at the time. Often, notes on lectures exist (prepared by the lecturer himself, by graduate students, or by postdoctoral fellows) and are distributed in mimeographed form on a limited basis. One of the principal purposes of the FRONTIERS IN PHYSICS Series is to make such notes available to a wider audience of physicists.

It should be emphasized that lecture notes are necessarily rough and informal, both in style and content; and those in the series will prove no exception. This is as it should be. The point of the series is to offer new, rapid, more informal, and, it is hoped, more effective ways for physicists to teach one another. The point is lost if only elegant notes qualify.

The publication of collections of reprints of recent articles in very active fields of physics will improve communication. Such collections are themselves useful to people working in the field. The value of the reprints will, however, be enhanced if the collection is accompanied by an introduction of moderate length which will serve to tie the collection together and, necessarily, constitute a brief survey of the present status of the field. Again, it is appropriate that such an introduction be informal, in keeping with the active character of the field.

The informal monograph, representing an intermediate step between lecture notes and formal monographs, offers an author the opportunity to present his views of a field which has developed to the point where a summation might prove extraordinarily fruitful but a formal monograph might not be feasible or desirable.

Contemporary classics constitute a particularly valuable approach to the teaching and learning of physics today. Here one thinks of fields that lie at the heart of much of present-day research, but whose essentials are by now well understood, such as quantum electrodynamics or magnetic resonance. In such fields some of the best pedagogical material is not readily available, either because it consists of papers long out of print or lectures that have never been published.

The above words written in August 1961 seem equally applicable today. The development during the past decade of a quantum theory of gauge fields represents a significant contribution to our understanding of elementary particles and their interactions.

The present volume is intended to introduce the reader to the methods of quantum gauge field theory; the authors both set forth the basic elements of the theory and provide illustrative applications.

Ludwig Faddeev and Andrei Slavnov are especially well qualified to write such a book because they have been among the key participants in the development of the theory.

It is a pleasure to welcome them to the ranks of contributors to this Series. I would like also to take this opportunity to thank Dr. Edward Witten of Harvard University for his invaluable assistance in reviewing and editing the English translation of this volume.

DAVID PINES

PREFACE TO THE ORIGINAL (RUSSIAN) EDITION

Progress in quantum field theory, during the last ten years, is to a great extent due to the development of the theory of Yang–Mills fields, sometimes called gauge fields. These fields open up new possibilities for the description of interactions of elementary particles in the framework of quantum field theory. Gauge fields are involved in most modern models of strong and also of weak and electromagnetic interactions. There also arise the extremely attractive prospects of unification of all the interactions into a single universal interaction.

At the same time the Yang–Mills fields have surely not been sufficiently considered in modern monographical literature. Although the Yang–Mills theory seems to be a rather special model from the point of view of general quantum field theory, it is extremely specific and the methods used in this theory are quite far from being traditional. The existing monograph of Konoplyova and Popov, "Gauge Fields," deals mainly with the geometrical aspects of the gauge field theory and illuminates the quantum theory of the Yang–Mills fields insufficiently. We hope that the present book to some extent will close this gap.

The main technical method, used in the quantum theory of gauge fields, is the path-integral method. Therefore, much attention is paid in this book to the description of this alternative approach to the quantum field theory. We have made an attempt to expound this method in a sufficiently self-consistent manner, proceeding from the fundamentals of quantum theory. Nevertheless, for a deeper understanding of the book it is desirable for the reader to be familiar with the traditional methods of quantum theory, for example, in the volume of the first four chapters of the book by N. N. Bogolubov and D. V. Shirkov, "Introduction to the Theory of Quantized Fields." In particular, we shall not go into details of comparing the Feynman diagrams to the terms of the perturbation-theory expansion, and of the rigorous substantiation of the renormalization procedure, based on the R-operation. These problems are not specific for the Yang–Mills theory and are presented in detail in the quoted monograph.

There are many publications on the Yang–Mills fields, and we shall not go into a detailed review of this literature to any extent. Our aim is to introduce the methods of the quantum Yang–Mills theory to the reader. We shall not discuss alternative approaches to this theory, but shall present in detail that approach, which seems to us the most simple and natural one. The applications dealt with in the book are illustrative in character and are not the last word to be said about applications of the Yang–Mills fields to elementary-particle models. We do this consciously, since the phenomenological aspects of gauge theories are developing and changing rapidly. At the same time the technique of quantization and renormalization of the Yang–Mills fields has already become well established. Our book is mainly dedicated to these specific problems.

We are grateful to our colleagues of the V. A. Steclov Mathematical Institute in Moscow and Leningrad for numerous helpful discussions of the problems dealt with in this book.

We would especially like to thank D. V. Shirkov and O. I. Zav'yalov who read the manuscript and made many useful comments, and E. Sh. Yegoryan for help in checking the formulas.

Moscow-Leningrad-Kirovsk L. D. FADDEEV, A. A. SLAVNOV

PREFACE TO THE ENGLISH EDITION

This book was written in the spring of 1977 and published in Russian in 1978. By that time, the perturbation theory for quantum Yang–Mills theory had been completed and its relevance for elementary particle theory also had been generally accepted. We hope that our book covers all essential features of this development.

Since publication of the Russian edition, several new and exciting ideas were proposed dealing mainly with the nonperturbative approach and the problem of quark confinement. In particular, the problem of gauge ambiguity, lattice formulation of gauge theories, the role of instanton solutions in quantum dynamics, have been widely discussed. However, it is our opinion that none of these approaches can be considered as definitive. For this reason, we decided that it would be premature to provide any additions to the present English-language version.

We are grateful to Dr. D. B. Pontecorvo for prompt and faithful translation into English.

Aspen, Colorado,
August, 1979

L. D. FADDEEV, A. A. SLAVNOV

GAUGE FIELDS

INTRODUCTION TO QUANTUM THEORY

CHAPTER 1

INTRODUCTION

1.1 BASIC CONCEPTS AND NOTATION

At first sight, the theory of gauge fields which we shall discuss in this book describes a rather narrow class of quantum-field-theory models. However, the opinion is becoming more and more popular that this theory has a chance to become the basis of the theory of elementary particles. This opinion is based on the following facts:

First, the only theory (quantum electrodynamics) completely confirmed by experiments is a particular case of the gauge theory.

Second, phenomenological models of weak interactions have acquired an elegant and self-consistent formulation in the framework of gauge theories. The phenomenological four-fermion interaction has been replaced by the interaction with an intermediate vector particle, the quantum of the Yang–Mills field. Existing experimental data along with the requirement of gauge invariance led to the prediction of weak neutral currents and of a new quantum number for hadrons (charm).

Third, it seems that phenomenological quark models of strong interactions also have their most natural foundation in the framework of gauge theories. Gauge theories give a unique possibility of describing, in the framework of quantum field theory, the phenomenon of asymptotic freedom. These theories also afford hopes of explaining quark confinement, although this question is not quite clear.

Finally, the extension of the gauge principle may lead to the gravitational interaction also being placed in the general scheme of Yang–Mills fields.

So the possibility arises of explaining, on the basis of one principle, all the hierarchy of interactions existing in nature. The term

L. D. Faddeev and A. A. Slavnov. Gauge Fields: Introduction to Quantum Theory, ISBN 0-8053-9016-2.

Copyright © 1980 by The Benjamin/Cummings Publishing Company, Advanced Book Program. All rights reserved. No part of this publication may be reproduced, stored in a retrieval system, or transmitted, in any form or by any means, electronic, mechanical photocopying, recording, or otherwise, without the prior permission of the publisher.

unified field theory, discredited some time ago, now acquires a new reality in the framework of gauge field theories. Independently of the question to what extent all these great expectations will be realized, the theory of Yang–Mills fields is today an essential method in theoretical physics and without doubt will have an important place in a future theory of elementary particles.

In the formation of this picture a number of scientists took part. Let us mention some of the key dates.

In 1953 Yang and Mills, for the first time, generalized the principle of gauge invariance of the interaction of electric charges for the case of interacting isospins. In their paper, they introduced a vector field, which later became known as the Yang–Mills field, and within the framework of the classical field theory its dynamics was developed.

In 1967 Faddeev and Popov, and de Witt, constructed a self-consistent scheme for the quantization of massless Yang–Mills fields. In the same year, Weinberg and Salam independently proposed a unified gauge model of weak and electromagnetic interactions, in which the electromagnetic field and the field of the intermediate vector boson were combined into a multiplet of Yang–Mills fields. This model was based on the mechanism of mass generation for vector bosons as a result of a spontaneous symmetry breaking, proposed earlier by Higgs and Kibble.

In 1971 G. 't Hooft showed that the general methods of quantization of massless Yang–Mills fields may be applied, practically without any change, to the case of spontaneously broken symmetry. Thus, the possibility was discovered of contructing a self-consistent quantum theory of massive vector fields, which are necessary for the theory of weak interactions and, in particular, for the Salam–Weinberg model.

By 1972 the construction of the quantum theory of gauge fields in the framework of perturbation theory was largely completed. In papers by A. Slavnov, by J. Taylor, by B. Lee and J. Zinn-Justin, and by G. 't Hooft and M. Veltman, various methods of invariant regularization were developed, the generalized Ward identities were obtained, and a renormalization procedure was constructed in the framework of the perturbation theory. This led to the construction of a finite and unitary scattering matrix for the Yang–Mills field.

Since then on, the theory of gauge fields has developed rapidly, both theoretically and phenomenologically. The history of this development may be illustrated by the rapporteurs' talks presented at international conferences on high-energy physics (B. Lee, 1972, Batavia; J. Illiopoulos, 1974, London; A. Slavnov, 1976, Tbilisi).

From the above short historical survey we shall pass on to the description of the Yang–Mills field itself. For this, we must first recal some notation from the theory of compact Lie groups. More specifically, we shall be interested mainly in the Lie algebras of these groups. Let Ω be a compact semisimple Lie group, that is, a compact group which has no invariant commutative (Abelian) subgroups. The number of independent parameters which characterize an arbitrary element of the group (that is, the dimension) is equal to n. Among the representations of this group and its Lie algebra, there exists the representation of $n \times n$ matrices (adjoint representation). It is generated by the natural action of the group on itself by the similarity transformations

$$h \to \omega h \omega^{-1}; \qquad h, \omega \in \Omega. \tag{1.1}$$

Any matrix $\mathcal{T}$ in the adjoint representation of the Lie algebra can be represented by a linear combination of n generators,

$$\mathcal{T} = T^a \alpha^a. \tag{1.2}$$

For us it is essential that the generators T^a can be normalized by the condition

$$\operatorname{tr}(T^a T^b) = -2\delta^{ab}. \tag{1.3}$$

In this case the structure constants t^{abc} which take part in the condition

$$[T^a, T^b] = t^{abc} T^c, \tag{1.4}$$

are completely antisymmetric. The reader unfamiliar with the theory of Lie groups may keep in mind just these two relationships, which are actually a characterizing property of the compact semisimple Lie group.

A compact semisimple group is called simple if it has no invariant Lie subgroups. A general semisimple group is a product of simple groups. This means that the matrices of the Lie algebra in the adjoint rerpresentation have a block-diagonal form, where each block corresponds to one of the simple factors. The generators of the group can be chosen so that each one has nonzero matrix elements only within one of the blocks. We shall always have in mind exactly such a choice of generators, in correspondence with the structure of the direct product.

The simplest example of such a group is the simple group $SU(2)$. The dimension of this group equals 3, the Lie algebra in the adjoint representation is given by the antisymmetric 3×3 matrices; as generators the matrices

$$T^1 = \begin{pmatrix} 0 & 0 & 0 \\ 0 & 0 & -1 \\ 0 & 1 & 0 \end{pmatrix}; \quad T^2 = \begin{pmatrix} 0 & 0 & 1 \\ 0 & 0 & 0 \\ -1 & 0 & 0 \end{pmatrix}; \quad T^3 = \begin{pmatrix} 0 & -1 & 0 \\ 1 & 0 & 0 \\ 0 & 0 & 0 \end{pmatrix}; \tag{1.5}$$

can be chosen; the structure constants t^{abc} in this base coincide with the completely antisymmetric tensor ε^{abc}.

Besides semisimple compact groups, we shall also deal with the commutative (Abelian) group $U(1)$. The elements of this group are complex numbers, with absolute values equal to unity. The Lie algebra of this group is one-dimensional and consists of imaginary numbers or of real antisymmetric 2×2 matrices.

The Yang–Mills field can be associated with any compact semisimple Lie group. It is given by the vector field $\mathscr{A}_\mu(x)$, with values in the Lie algebra of this group. It is convenient to consider $\mathscr{A}_\mu(x)$ to be a matrix in the adjoint representation of this algebra. In this case it is defined by its coefficients $A_\mu^a(x)$:

$$\mathscr{A}_\mu(x) = A_\mu^a(x)\, T^a \tag{1.6}$$

with respect to the base of the generators T^a.

In the case of the group $U(1)$ the electromagnetic field $\mathscr{A}_\mu(x) = i\,A_\mu(x)$ is an analogous object.

We shall now pass on to the definition of the gauge group and its action on Yang–Mills fields. In the case of electrodynamics the gauge transformation is actually the well-known gradient transformation

$$\mathscr{A}_\mu(x) \to \mathscr{A}_\mu(x) + i\partial_\mu\lambda(x). \tag{1.7}$$

Let us recall its origin in the framework of the classical field theory. The electromagnetic field interacts with charged fields, which are described by complex functions $\psi(x)$. In the equations of motion the field $\mathscr{A}_\mu(x)$ always appears in the following combination:

$$\nabla_\mu\psi = (\partial_\mu - \mathscr{A}_\mu)\,\psi = (\partial_\mu - iA_\mu)\,\psi. \tag{1.8}$$

The above gradient transformation provides the covariance of this combination with respect to the phase transformation of the fields ψ. If ψ transforms according to the rule

$$\begin{aligned} \psi(x) &\to e^{i\lambda(x)}\psi(x), \\ \bar{\psi}(x) &\to e^{-i\lambda(x)}\bar{\psi}(x), \end{aligned} \tag{1.9}$$

then $\nabla_\mu \psi$ transforms in the same way. Indeed,

$$\begin{aligned} (\partial_\mu - \mathscr{A}_\mu)\psi \to [\partial_\mu - i\partial_\mu\lambda(x) - \mathscr{A}_\mu(x)]\, e^{i\lambda(x)}\psi(x) = \\ = e^{i\lambda(x)}[\partial_\mu - \mathscr{A}_\mu(x)]\,\psi(x). \end{aligned} \tag{1.10}$$

As a result, the equations of motion are also covariant with respect to the transformations (1.7) and (1.9); if the pair $\psi(x)$, $\mathscr{A}_\mu(x)$ is a solution, the $e^{i\lambda(x)}\psi(x)$, $\mathscr{A}_\mu(x) + i\partial_\mu\lambda(x)$ is also a solution.

In other words, a local change in phase of the field $\psi(x)$, which can be considered to be the coordinate in the charge space, is equivalent to the appearance of an additional electromagnetic field. We see here a complete analogy with the weak equivalence principle in Einstein's theory of gravity, where a change of the coordinate system leads to the appearance of an additional gravitational field.

Extending this analogy further, one may formulate the relativity principle in the charge space. This principle was first introduced by Hermann Weyl in 1929: The field configurations $\psi(x)$, $\mathscr{A}_\mu(x)$and $\psi(x)e^{i\lambda(x)}$, $\mathscr{A}_\mu(x) + i\partial_\mu\lambda(x)$ describe the same physical situation. If the construction of theory is based on this principle, then the above-described way of constructing the equations of motion in terms of covariant derivatives is the only possible one.

The generalization of this principle to the case of the more complicated charge space leads to the Yang–Mills theory. Examples of such charge (or internal, as they are often called) spaces are the isotopic space, the unitary-spin space in the theory of hadrons, and so on. In all these examples we deal with fields $\psi(x)$ that acquire values in the charge space, which itself is a representation space for some compact semisimple groups Ω [$SU(2)$, $SU(3)$, etc.]. The equations of motion for the fields $\psi(x)$ contain the covariant derivative

$$\nabla_\mu = \partial_\mu - \Gamma(\mathscr{A}_\mu), \tag{1.11}$$

where $\Gamma(\mathscr{A}_\mu)$ is the representation of the matrix $\mathscr{A}_\mu$ corresponding to the given representation of the group Ω. For example, if $\Omega = SU(2)$, and the charge space corresponds to the two-dimensional representation, then the above-mentioned generators T^a are represented by the Pauli matrices

$$\Gamma(T^a) = \frac{1}{2i}\tau^a, \tag{1.12}$$

where

$$\tau^1 = \begin{pmatrix} 0 & 1 \\ 1 & 0 \end{pmatrix}, \quad \tau^2 = \begin{pmatrix} 0 & -i \\ i & 0 \end{pmatrix}, \quad \tau^3 = \begin{pmatrix} 1 & 0 \\ 0 & -1 \end{pmatrix}, \tag{1.13}$$

and in this case

$$\Gamma(\mathcal{A}_\mu) = \frac{1}{2i} A^a_\mu \tau^a. \tag{1.14}$$

The transformation of the fields $\psi(x)$ analogous to the local phase transformation in electrodynamics has the following form:

$$\psi(x) \to \psi^\omega(x) = \Gamma[\omega(x)]\,\psi(x), \tag{1.15}$$

where $\omega(x)$ is a function of x which has its values in the group Ω. It is convenient to consider $\omega(x)$ to be a matrix in the adjoint representation of the group Ω. The derivative (1.11) will be covariant with respect to this transformation if the field $\mathcal{A}_\mu(x)$ transforms according to the rule

$$\mathcal{A}_\mu(x) \to \mathcal{A}^\omega_\mu(x) = \omega(x)\,\mathcal{A}_\mu(x)\,\omega^{-1}(x) + \partial_\mu\omega(x)\,\omega^{-1}(x). \tag{1.16}$$

It is not difficult to see that this transformation obeys the group law. The set of these transformations composes a group which may formally be denoted by

$$\tilde{\Omega} = \prod_x \Omega. \tag{1.17}$$

This group is called the group of gauge transformations.

Often it is convenient to deal with the infinitesimal form of the gauge transformation. Let the matrices $\omega(x)$ differ infinitesimally from the unit matrix

$$\omega(x) = 1 + \alpha(x) = 1 + \alpha^a(x)\,T^a, \tag{1.18}$$

where $\alpha(x)$ belongs to the Lie algebra of the group Ω. Then the change of $\mathscr{A}_\mu$ under such a transformation will be

$$\delta\mathscr{A}_\mu = \partial_\mu \alpha - [\mathscr{A}_\mu, \alpha] = \nabla_\mu \alpha, \tag{1.19}$$

or for the components,

$$\delta A_\mu^a = \partial_\mu \alpha^a - t^{abc} A_\mu^b \alpha^c. \tag{1.20}$$

The corresponding transformation for ψ takes the form

$$\delta\psi = \Gamma(\alpha)\psi. \tag{1.21}$$

It is obvious that the group of gradient transformations in electrodynamics is a particular case of the gauge group.

The existence of covariant derivatives makes it possible to dynamically realize the relativity principle in the internal space: The field configurations $\psi(x)$, $\mathscr{A}_\mu(x)$ and $\Gamma[\omega(x)]\psi(x)$, $\mathscr{A}_\mu^\omega(x)$ describe the same physical situation. If we take this principle as a basis for constructing the dynamics, we then automatically come to the Yang–Mills theory.

The relativity principle means that not only one set of fields, but a whole class of gauge-equivalent configurations corresponds to the true physical configuration. To be clearer, this principle implies that in the internal charge space there is no special fixed basis with respect to which the physical fields of matter ψ are represented in terms of components: $\psi = (\psi_1, \ldots, \psi_m)$. Such a basis can be introduced locally at each space-time point; however, there is no physical reason for fixing its position. The local change of basis is interpreted as a change of the gauge field, which plays a role analogous to the role of gravitational or electromagnetic fields.

The relativity principle leads to a significant formal difference in the description of the dynamics of gauge fields in comparison with more customary fields such as, for example, the self-interacting scalar field. In order to work in practice with classes of equivalent configurations, they must somehow be parametrized, that is, in each class unique representatives must be chosen. Usually, this is achieved by imposing a subsidiary condition which eliminates the gauge freedom. This subsidiary condition is called the gauge condition, or simply gauge. The most frequently used gauges are the following conditions:

$$\begin{aligned} \Phi_L &\equiv \partial_\mu \mathscr{A}_\mu = 0 \quad \text{(Lorentz gauge)}, \\ \Phi_C &\equiv \partial_k \mathscr{A}_k = 0 \quad \text{(Coulomb gauge)}, \\ \Phi_H &\equiv \mathscr{A}_0 = 0 \quad \text{(Hamilton gauge)}, \\ \Phi_A &\equiv \mathscr{A}_3 = 0 \quad \text{(axial gauge)}. \end{aligned} \tag{1.22}$$

For a general system including both fields $\mathscr{A}_\mu$ and fields ψ, the latter may enter into the gauge condition. Examples of such conditions will be presented in Section 1.3.

In general the gauge condition $\Phi(A, \psi; x)$ is a family of functionals of $\mathscr{A}_\mu$ and ψ, one for each x. For fixed x, $\Phi(A, \psi; x)$ is an element of the Lie algebra of the group G, so the number of independent gauge conditions coincides with the dimension of the gauge group. In the examples (1.22) all the conditions are exactly of such a form. Furthermore, in these examples the gauge conditions are local, that is $\Phi(A, \psi; x)$ depends on the values of $\mathscr{A}_\mu$ and ψ in the neighborhood of point x.

Let us discuss the requirements to be satisfied by the gauge conditions. The most important one implies that the system of equations

$$\Phi(A^\omega, \psi^\omega; x) = 0 \tag{1.23}$$

has a unique solution $\omega(x)$ for fixed $\mathscr{A}_\mu$ and ψ. This requirement means that in each set of equivalent fields there actually exists a unique set of fields $\mathscr{A}_\mu$, ψ which satisfies the condition (1.23). This set, considered as a representative of the class, characterizes uniquely the true physical configuration. Another requirement that is less fundamental, although important practically, is that Equation (1.23) must not be too complicated and should give a sufficiently explicit solution $\omega(x)$, at least in the framework of perturbation theory.

Equation (1.23) is a system of nonlinear equations for $\omega(x)$. For local gauge conditions it is a nonlinear system of partial differential equations. For instance, for the Lorentz gauge this system of equations takes the following form:

$$\begin{aligned} \nabla_\mu L_\mu &= \partial_\mu L_\mu - [\mathscr{A}_\mu, L_\mu] = -\partial_\mu \mathscr{A}_\mu; \\ L_\mu &= \omega^{-1} \partial_\mu \omega, \end{aligned} \tag{1.24}$$

and for small $\mathscr{A}_\mu$ and $\alpha(x)$ it is rewritten as

$$\Box\,\alpha - [\mathscr{A}_\mu, \partial_\mu \alpha] + \ \ldots\ = -\partial_\mu \mathscr{A}_\mu, \qquad (1.25)$$

where the dots stand for terms of higher order in α. Equation (1.25) can be uniquely solved with respect to α in the framework of perturbation theory if the operator $\Box = \partial_\mu \partial_\mu$ is supplied with suitable boundary conditions. Such boundary conditions arise in the description of the dynamics of Yang–Mills fields and will be discussed in Chapter Three. Nevertheless, beyond the domain of perturbation theory for large fields $\mathscr{A}_\mu$, the uniqueness of the solution of equation (1.24) may fail. Discussion of this possibility is not within the scope of this book.

A necessary condition for the solvability of the equations (1.23) is the nondegeneracy of the corresponding Jacobian. Variation of the gauge condition under an infinitesimal gauge transformation of α defines the linear operator M_Φ which acts on α:

$$M_\Phi \alpha = \int \Big[\frac{\delta \Phi(A, \psi; x)}{\delta \mathscr{A}_\mu(y)} (\partial_\mu \alpha(y) - [\mathscr{A}_\mu(y), \alpha(y)]) + \\ + \frac{\delta \Phi(A, \psi; x)}{\delta \psi(y)} \Gamma(\alpha(y)) \psi(y) \Big] dy, \qquad (1.26)$$

This operator plays the role of the Jacobian matrix for the condition (1.23). Nondegeneracy of the operator M_Φ,

$$\det M_\Phi \neq 0 \qquad (1.27)$$

is a necessary condition for the existence of a unique solution for the system (1.23).

For local gauge conditions M_Φ is a differential operator which is obtained while linearizing the system (1.23). For example, in the case of the Lorentz gauge condition, $M_{\Phi_L} = M_L$ has the form

$$M_L \alpha = \Box\,\alpha - \partial_\mu [\mathscr{A}_\mu, \alpha].$$

This operator is uniquely reversible within the framework of perturbation theory provided boundary conditions are introduced. As was noted above, these conditions will be discussed in Chapter Three.

We shall call the condition (1.27) the admissibility condition for the gauge condition, and it will be frequently discussed further on.

1.2. GEOMETRICAL INTERPRETATION OF THE YANG–MILLS FIELD

The construction described in the previous section allows an elegant geometrical interpretation when the Yang–Mills fields play the samc role as the Christoffel symbols in gravitation theory. Analogously to the latter, the Yang–Mills fields describe parallel translation in the charge space and determine the curvature of this space, the fields $\psi(x)$ being analogs of tensor fields.

A natural geometrical language for the description of this analogy is yielded by the fiber-bundle theory. In this theory the concept of connection in the principle bundle corresponds to the Yang–Mills field. Although the fiber-bundle theory produces the most adequate language for axiomatization of the classical field theory, in this book, which is addressed mainly to physicists, we shall not use it. We just point out that the general concept of connection, equivalent to the Yang–Mills field, appeared in the mathematical literature only in 1950, that is, practically simultaneously with the work of Yang and Mills.

Let us now explain in what sense the Yang–Mills fields determine parallel translation. Let $\gamma(s)$ be a contour in space-time defined by the equation

$$x_\mu = x_\mu(s). \tag{2.1}$$

The vector field $\dot{\gamma}(s)$ with components

$$X_\mu = \frac{dx_\mu}{ds} \tag{2.2}$$

is tangential to the curve $\gamma(s)$ at each of its points. We shall say that the field $\psi(x)$ undergoes parallel translations along $\gamma(s)$ if at each point of the contour,

$$\nabla_\mu \psi(x)\,|_{x=x(s)}\, X_\mu = 0, \tag{2.3}$$

that is, the covariant derivative in the tangential direction is equal to zero.

Generally speaking, parallel translation along a closed contour changes the field $\psi(x)$. Let us calculate this change for an infinitesimal contour. We shall consider a contour which has the form of a parallelogram with vertices

$$(x,\ x + \Delta_1 x,\ x + \Delta_1 x + \Delta_2 x,\ x + \Delta_2 x).$$

It may be readily verified that if the covariant derivative along this contour is equal to zero, then the total change in $\psi(x)$ corresponding to a whole turn round the closed contour is equal to

$$\Delta_{12}\psi(x) = \Gamma(\mathcal{F}_{\mu\nu})\,\psi\,(\Delta_1 x_\mu \Delta_2 x_\nu - \Delta_1 x_\nu \Delta_2 x_\mu), \tag{2.4}$$

where

$$\mathcal{F}_{\mu\nu} = \partial_\nu \mathcal{A}_\mu - \partial_\mu \mathcal{A}_\nu + [\mathcal{A}_\mu, \mathcal{A}_\nu]. \tag{2.5}$$

Indeed, since the covariant derivative along the side $(x, x + \Delta_1 x)$ is equal to zero, the change in $\psi(x)$ corresponding to the change of x along the first contour equals

$$\Delta_1\psi(x) = \psi(x + \Delta_1 x) - \psi(x) = \partial_\mu \psi \Delta_1 x_\mu = \Gamma(\mathcal{A}_\mu)\,\psi(x)\,\Delta_1 x_\mu. \tag{2.6}$$

Performing analogous calculations for the remaining sides of the parallelogram, and taking into account the linear dependence of $\Gamma(\mathcal{A}_\mu)$ upon $\mathcal{A}_\mu$ and the fact that $[\Gamma(\mathcal{A}_\mu), \Gamma(\mathcal{A}_\nu)] = \Gamma([\mathcal{A}_\mu, \mathcal{A}_\nu])$, we obtain the formula (2.4) for the total change in $\psi(x)$. This formula shows that it is natural to call $\mathcal{F}_{\mu\nu}$ the curvature of the charge space. Under gauge transformations $\Delta\psi(x)$ changes in the same way as $\psi(x)$. This is because for the construction of $\Delta\psi(x)$ we have used only the covariant derivative. Then from (2.4) it follows that $\Gamma(\mathcal{F}_{\mu\nu}(x))$ transforms according to the law

$$\Gamma(\mathcal{F}_{\mu\nu}(x)) \to \Gamma(\omega(x))\,\Gamma(\mathcal{F}_{\mu\nu}(x))\,\Gamma(\omega^{-1}(x)). \tag{2.7}$$

Therefore $\mathcal{F}_{\mu\nu}(x)$ itself under gauge transformations transforms as

$$\mathcal{F}_{\mu\nu}(x) \to \omega(x)\,\mathcal{F}_{\mu\nu}(x)\,\omega^{-1}(x). \tag{2.8}$$

If we adopt the convention that $\psi(x)$ is a vector with respect to gauge transformations, the $\Gamma(\mathcal{F}_{\mu\nu}(x))$ is a tensor of rank two. And $\mathcal{F}_{\mu\nu}(x)$ itself is sometimes conveniently considered a vector in the adjoint representation.

Our indirect derivation of (2.8) is verified by a straightforward check if one takes advantage of the explicit expression (2.5) for $\mathcal{F}_{\mu\nu}(x)$ in terms of $\mathcal{A}_\mu(x)$ and of the transformation law (1.16) for

$\mathcal{A}_\mu(x)$. Thus we conclude our short description of the geometrical interpretation of the Yang–Mills fields: They describe parallel translation of vectors in the charge space, and the tensor $\mathcal{F}_{\mu\nu}(x)$ is the curvature tensor of this space. The reader familiar with the theory of gravity must surely have already noticed the complete analogy between $\mathcal{A}_\mu(x)$ and Christoffel symbols, and between $\mathcal{F}_{\mu\nu}(x)$ and the curvature tensor of the gravitational field. To conclude this analogy, we point out that the tensor $\mathcal{F}_{\mu\nu}(x)$ is the commutator of the covariant derivatives

$$\mathcal{F}_{\mu\nu}(x) = [\nabla_\mu, \nabla_\nu] \tag{2.9}$$

and the Jacobi identity

$$[[\nabla_\mu, \nabla_\nu], \nabla_\sigma] + \text{cyclic permutations} = 0 \tag{2.10}$$

leads to the identity

$$\nabla_\sigma \mathcal{F}_{\mu\nu}(x) + \text{cyclic permutations} = 0, \tag{2.11}$$

where $\nabla_\sigma \mathcal{F}_{\mu\nu}(x) = \partial_\sigma \mathcal{F}_{\mu\nu}(x) - [\mathcal{A}_\sigma(x), \mathcal{F}_{\mu\nu}(x)]$, which is the analog of the Bianchi identity in the theory of gravity. A similar consideration can be carried out in the case of the Abelian group $U(1)$. In this case

$$\mathcal{F}_{\mu\nu}(x) = \partial_\nu \mathcal{A}_\mu(x) - \partial_\mu \mathcal{A}_\nu(x) = i(\partial_\nu A_\mu(x) - \partial_\mu A_\nu(x)), \tag{2.12}$$

which obviously coincides with the tensor of the electromagnetic field strength. The interpretation of $\mathcal{F}_{\mu\nu}(x)$ as the curvature of the charge space, originating with Fock and Hermann Weyl, is the most natural approach to the geometrization of the electromagnetic field. Numerous attempts to relate this field to the geometrical properties of space-time itself have never had any success.

In concluding this section we shall say a few words about the classical dynamics of the Yang–Mills field. Our task is to construct a gauge-invariant Lagrange function coinciding, in the case of the Abelian group $U(1)$, with the Lagrangian of the electromagnetic field

$$\mathcal{L} = \frac{1}{4e^2} \mathcal{F}_{\mu\nu} \mathcal{F}_{\mu\nu} + \mathcal{L}_M(\psi, \nabla_\mu \psi), \tag{2.13}$$

where $\mathcal{L}_M$ describes the gauge-invariant interaction of the fields $\mathcal{A}_\mu(x)$ and $\psi(x)$ and is deduced from the free Lagrangian of the fields ψ by replacing ordinary derivatives with covariant ones, and e plays the role of the electric charge. This formula may be easily rewritten in a more familiar form if one changes the normalization of the fields:

$$\mathcal{A}_\mu(x) \to e\mathcal{A}_\mu(x). \tag{2.14}$$

In this case the factor e^{-2} vanishes from the first term, but appears instead in the expression for the covariant derivative,

$$\nabla_\mu \to \partial_\mu - e\mathcal{A}_\mu.$$

In the following we shall use both methods of normalizing the fields $\mathcal{A}_\mu(x)$, without specially mentioning this.

A natural (and the only possible) generalization of the formula (2.13) to the case of the simple non-Abelian gauge group is the following expression:

$$\mathcal{L} = \frac{1}{8g^2}\operatorname{tr}\mathcal{F}_{\mu\nu}\mathcal{F}_{\mu\nu} + \mathcal{L}_M(\psi, \nabla_\mu\psi). \tag{2.15}$$

The first term may be rewritten also as

$$\mathcal{L} = -\frac{1}{4g^2}F^a_{\mu\nu}F^a_{\mu\nu}, \tag{2.16}$$

where $F^a_{\mu\nu}(x)$ are the components of the matrix $\mathcal{F}_{\mu\nu}(x)$ with respect to the base T^a. Obviously, this Lagrangian is invariant with respect to the gauge transformations (1.15), (1.16). In the case of the semisimple group of general form, the Lagrangian contains r arbitrary constants g_i, $i = 1,\ldots,r$, where r is the number of invariant simple factors. Then the formula analogous to (2.16) takes the form

$$\mathcal{L} = \sum_i -\frac{1}{4g_i^2}F^{a_i}_{\mu\nu}F^{a_i}_{\mu\nu}, \tag{2.17}$$

where i is the index number of a simple factor.

Contrary to electrodynamics, the Lagrangian (2.16) of the Yang–Mills field in vacuum (that is, in the absence of the fields ψ), in addition to the second-order terms in the fields, contains also higher-order terms. This means that Yang–Mills fields have nontrivial self-interaction. In other words, quanta of the Yang–Mills field themselves have charges, the interaction of which they transfer. The main specific feature of the Yang–Mills field dynamics is related to this self-action; therefore we shall often confine ourselves to the model of the Yang–Mills field in vacuum when dealing with general problems.

The equations of motion arising from the Lagrangian (2.16) for the Yang–Mills field in vacuum have the form

$$\nabla_\mu \mathcal{F}_{\mu\nu} = \partial_\mu \mathcal{F}_{\mu\nu} - [\mathcal{A}_\mu, \mathcal{F}_{\mu\nu}] = 0, \tag{2.18}$$

and written in terms of the $\mathcal{A}_\mu$,

$$\Box \mathcal{A}_\nu - \partial_\nu \partial_\mu \mathcal{A}_\mu + [\mathcal{A}_\mu, (\partial_\nu \mathcal{A}_\mu - \partial_\mu \mathcal{A}_\nu + [\mathcal{A}_\mu, \mathcal{A}_\nu])] - \\ - \partial_\mu [\mathcal{A}_\mu, \mathcal{A}_\nu] = 0, \tag{2.19}$$

represent a system of second-order equations. These equations are gauge-invariant in the following sense: If $\mathcal{A}_\mu$ is a solution of (2.19), then $\mathcal{A}_\mu^\omega$ is also a solution for any arbitrary $\omega(x)$. This means that the standard parametrization of the solutions in terms of the initial conditions [$\mathcal{A}_\mu(x,t)$, $\partial_0 \mathcal{A}_\mu(x,t)$ at a fixed t] is unfit for the system (2.19). By imposing the gauge conditions this obstacle is eliminated; however, the initial conditions are then not arbitrary, but are restricted by the gauge conditions.

Models of interaction of the Yang–Mills field with fields of matter will be considered in the next section.

1.3. DYNAMICAL MODELS WITH GAUGE FIELDS

The Lagrangian describing the interaction of the Yang–Mills field with spinor fields is the simplest one. Let a multiplet of spinor fields $\psi_k(x)$ realize the representation $\Gamma(\omega)$ of a simple compact gauge group Ω. Then the Lagrangian has the form

$$\mathcal{L} = \mathcal{L}_{YM} + i\bar{\psi}(x)\gamma_\mu \nabla_\mu \psi(x) - m\bar{\psi}(x)\psi(x). \tag{3.1}$$

We have used here the following notation: $\mathcal{L}_{YM}$ is the already familiar Lagrangian of the Yang–Mills field in vacuum,

$$\mathcal{L}_{YM} = \frac{1}{8g^2} \operatorname{tr} \mathcal{F}_{\mu\nu} \mathcal{F}_{\mu\nu}. \tag{3.2}$$

In the scalar product of two spinors the sum is over the indices corresponding to internal degrees of freedom; for instance, the mass term may be written down as

$$m\bar{\psi}(x)\,\psi(x) = m\bar{\psi}_k(x)\,\psi_k(x). \tag{3.3}$$

Furthermore,

$$(\nabla_\mu \psi(x))_k = \partial_\mu \psi_k(x) - (\Gamma(\mathscr{A}_\mu(x)))_{kl}\,\psi_l(x), \tag{3.4}$$

where $(\Gamma(\mathscr{A}_\mu))_{kl} = A_\mu^a (\Gamma(T^a))_{kl}$, and the matrix $(\Gamma(T^a))_{kl}$, which in the following will be denoted simply by Γ^α_{kl}, is the matrix of the generator T^α in the representation realized by the fields $\psi(x)$. Then

$$\bar{\psi}(x)\,\gamma_\mu \nabla_\mu \psi(x) = \bar{\psi}_k(x)\,\gamma_\mu \left(\partial_\mu \psi_k(x) - A^a_\mu(x)\,\Gamma^a_{kl}\psi_l(x)\right). \tag{3.5}$$

For example, let the gauge group be $\Omega = SU(2)$, and the fields $\psi(x)$ realize the fundamental representation of this group. Then

$$(\Gamma(\mathscr{A}_\mu))_{kl} = -\frac{i}{2} A^a_\mu (\tau^a)_{kl}, \tag{3.6}$$

where τ^α are the Pauli matrices, and the complete Lagrangian has the form

$$\mathscr{L} = -\frac{1}{4g^2}(\partial_\nu A^a_\mu - \partial_\mu A^a_\nu + \varepsilon^{abc} A^b_\mu A^c_\nu)^2 + \\ + i\bar{\psi}\gamma_\mu\left(\partial_\mu \psi + \frac{i}{2} A^a_\mu \tau^a \psi\right) - m\bar{\psi}\psi. \tag{3.7}$$

In the case when the gauge group is the group $SU(3)$ and the spinors $\psi(x)$ realize its fundamental (spinor) representation, the analogous Lagrangian takes the form

$$\mathscr{L} = -\frac{1}{4g^2}(\partial_\nu A^a_\mu - \partial_\mu A_\nu + f^{abc} A^b_\mu A^c_\nu)^2 + \\ + i\bar{\psi}\gamma_\mu\left(\partial_\mu \psi + \frac{i}{2} A^a_\mu \lambda^a \psi\right) - m\bar{\psi}\psi, \tag{3.8}$$

where f^{abc} are the structure constants of the group $SU(3)$, and the matrices λ^a are the well-known Gell-Mann matrices:

$$\lambda_1 = \begin{pmatrix} 0 & 1 & 0 \\ 1 & 0 & 0 \\ 0 & 0 & 0 \end{pmatrix}; \quad \lambda_2 = \begin{pmatrix} 0 & -i & 0 \\ i & 0 & 0 \\ 0 & 0 & 0 \end{pmatrix}; \quad \lambda_3 = \begin{pmatrix} 1 & 0 & 0 \\ 0 & -1 & 0 \\ 0 & 0 & 0 \end{pmatrix};$$

$$\lambda_4 = \begin{pmatrix} 0 & 0 & 1 \\ 0 & 0 & 0 \\ 1 & 0 & 0 \end{pmatrix}; \quad \lambda_5 = \begin{pmatrix} 0 & 0 & -i \\ 0 & 0 & 0 \\ i & 0 & 0 \end{pmatrix}; \quad \lambda_6 = \begin{pmatrix} 0 & 0 & 0 \\ 0 & 0 & 1 \\ 0 & 1 & 0 \end{pmatrix};$$

$$\lambda_7 = \begin{pmatrix} 0 & 0 & 0 \\ 0 & 0 & -i \\ 0 & i & 0 \end{pmatrix}; \quad \lambda_8 = \frac{1}{\sqrt{3}} \begin{pmatrix} 1 & 0 & 0 \\ 0 & 1 & 0 \\ 0 & 0 & -2 \end{pmatrix}. \tag{3.9}$$

Renormalization of the fields

$$A_\mu^a(x) \to g A_\mu^a(x) \tag{3.10}$$

changes the form of Lagrangians (3.7) and (3.8) to a more familiar one, the g being involved only in the interaction term.

The latter Lagrangian is used, for example, in the theory of strong interactions. In this case the spinors ψ are identified with the quark fields, the Yang–Mills fields are called "gluons", and the internal space is called the space of colors.

In the above-considered examples, when the gauge group is simple, all interactions are characterized by a single coupling constant. Such universality of the interactions is a specific feature of the Yang–Mills theory.

The next useful example is the interaction of the Yang–Mills field with a scalar field. Let the multiplet of scalar fields $\varphi_k(x)$ realize a real representation $\Gamma(\omega)$ of the simple compact group Ω. Then the gauge-invariant Lagrangian has the form

$$\mathcal{L} = \mathcal{L}_{YM} + \frac{1}{2} \nabla_\mu \varphi \nabla_\mu \varphi - \frac{m^2}{2} \varphi\varphi - V(\varphi), \tag{3.11}$$

where the covariant derivative $\nabla_\mu \varphi$ is constructed as above

$$\nabla_\mu \varphi = \partial_\mu \varphi - \Gamma(\mathcal{A}_\mu)\varphi, \tag{3.12}$$

$\varphi\varphi$, as before, is a scalar product in the charge space, and $V(\varphi)$ is a fôrm of third and fourth degree in the fields that is invariant with respect to the group Ω.

In the case when $\Omega = SU(2)$ and fields φ realize the adjoint representation $\varphi = \varphi^a$, $a = 1, 2, 3$, the corresponding formula becomes

$$\mathcal{L} = \mathcal{L}_{YM} + \frac{1}{2} (\partial_\mu \varphi^a - g\varepsilon^{abc} A_\mu^b \varphi^c)^2 - \frac{m^2}{2} \varphi^a \varphi^a - \lambda^2 (\varphi^a \varphi^a)^2, \tag{3.13}$$

where parameters m and λ^2 play the role of the masses and of the contact-interaction coupling constants of the scalar fields. The Lagrangian (3.13) itself is, evidently, of little interest from the viewpoint of physical applications; however, an insignificant-looking modification leads to the extremely interesting possibility of describing massive vector fields within the framework of the Yang–Mills theory. This mechanism for the mass generation of the vector field is called the Higgs effect. We now proceed to discuss this effect.

We shall continue to deal with the gauge group $SU(2)$ as an example. We first consider the case when the scalar field belongs to the adjoint representation. We shall use the following Lagrangian:

$$\mathscr{L} = \mathscr{L}_{YM} + \frac{1}{2}(\nabla_\mu \varphi^a)^2 - \lambda^2 (\varphi^a \varphi^a - \mu^2)^2. \qquad (3.14)$$

This Lagrangian differs from the Lagrangian (3.13) that we examined earlier by the constant term $-\lambda^2\mu^4$ and the sign of the term with φ squared. At first sight, the Lagrangian (3.14) describes particles with imaginary masses, and therefore has no physical meaning. Such a conclusion, however, would be too hasty. The term with φ squared plays the role of the mass only if $\varphi = 0$ is the stable equilibrium point and is, therefore, the potential-energy minimum. In our case the potential energy is

$$U(A_\mu, \varphi) = \int \left[\frac{1}{4g^2} F^a_{ik} F^a_{ik} + \frac{1}{2}\nabla_i \varphi^a \nabla_i \varphi^a + \lambda^2 (\varphi^a \varphi^a - \mu^2)^2\right] d^3x,$$
$$i, k = 1, 2, 3, \qquad (3.15)$$

and the configuration $\varphi^a = 0$, $A^a_\mu = 0$ is a saddle point. The corresponding equilibrium is unstable. However, there exist also stable equilibrium points; they are the configurations corresponding to zero A^a_μ and to constant φ with a fixed length $\varphi^2 = \mu^2$. Such $\mathscr{A}_\mu$, φ nullify all the three positive terms which make up the potential energy. (It should be pointed out that besides these configurations themselves, gauge transformation of these configurations obviously yields configurations which are also minima. However, owing to the relativity principle, these configurations present no new physical information, and we shall not take them into consideration.)

Besides these translation-invariant minima the potential energy has also other ones, for example, minima corresponding to the 't Hooft–Polyakov monopoles. The energy values for these configurations, however, are higher, so they are only local minima.

In order to determine the real masses, it is necessary to expand the potential energy in Taylor's series around the true minimum. In our case the equilibrium point is degenerate. The minimal configurations form a two-dimensional sphere S^2 with points corresponding to directions of the constant vector φ. We shall denote these directions by $\boldsymbol{n}$ and write the corresponding φ with the index n, so that $\boldsymbol{\varphi}_n = \mu\boldsymbol{n}$. The degeneracy is eliminated if we reduce the configuration space and take into consideration only fields φ which coincide asymptotically with one of the φ_n at high $|x|$. Such a choice, naturally, destroys the invariance under $SU(2)$ transformations with constant parameters (isotopic invariance). It may be shown that this condition does not contradict the dynamics, and that theories corresponding to different choices of φ_n are physically equivalent. The reader familiar with the solid-state physics will, of course, see here an analogy to the theory of ferromagnetics, in which a choice of the direction of the spontaneous magnetization must be made in order to formulate the theory itself.

For definiteness, let us choose the vector $\boldsymbol{n}$ to be directed along the third axis: $\boldsymbol{n} = (0, 0, 1)$. The corresponding vector $\boldsymbol{\varphi}_n$ is $(0,0,\mu)$.

Transition to fields $\varphi(x)$ with a zero asymptote at infinity,

$$\varphi(x) \to \varphi_n + \varphi(x) \tag{3.16}$$

makes the isotopic-symmetry breaking explicit, and the Lagrangian takes the form

$$\begin{aligned} \mathscr{L} = \mathscr{L}_{YM} + \frac{1}{2}(\nabla_\mu\varphi^a)^2 + \frac{m_1^2}{2}[(A_\mu^1)^2 + (A_\mu^2)^2] + \\ + m_1(A_\mu^1\partial_\mu\varphi^2 - A_\mu^2\partial_\mu\varphi^1) + gm_1[\varphi^3[(A_\mu^1)^2 + (A_\mu^2)^2] - \\ - A_\mu^3[\varphi^1 A_\mu^1 + \varphi^2 A_\mu^2]] - \frac{m_2^2}{2}(\varphi^3)^2 - \frac{m_2^2 g}{2m_1}\varphi^3(\varphi^a)^2 - \frac{m_2^2 g^2}{8m_1^2}(\varphi^a\varphi^a)^2, \\ m_1 = \mu g; \quad m_2 = 2\sqrt{2}\,\lambda\mu. \end{aligned} \tag{3.17}$$

Although we have explicitly broken isotopic invariance, the Lagrangian and the boundary conditions are invariant under local gauge transformations with functions $\omega(x)$ tending to unity at infinity. We shall give the explicit form of the gauge transformations in new variables, confining ourselves to infinitesimal transformations:

$$\delta\varphi^a(x) = -g\varepsilon^{abc}\varphi^b(x)\alpha^c(x) - m_1\varepsilon^{a3c}\alpha^c(x). \tag{3.18}$$

In order to analyze the spectrum of masses generated by the Lagrangian (3.17), one must choose representatives in gauge-equivalent classes of fields $\mathscr{A}_\mu(x)$, $\varphi(x)$, that is, one must fix the gauge. It is convenient to choose the following gauge condition:

$$\varphi^1(x)=0; \quad \varphi^2(x)=0; \quad \partial_\mu A^3_\mu(x)=0. \tag{3.19}$$

It can be verified that for sufficiently small $\varphi^3(x)$ the admissibility condition is fulfilled. Indeed,

$$\delta(\partial_\mu A^3_\mu)=\Box\,\alpha^3-g\varepsilon^{3bc}\partial_\mu[A^b_\mu\alpha^c] \tag{3.20}$$

and $\delta\varphi^{1,2}$ are determined by the formula (3.18). As a result, the operator M corresponding to our gauge has the form

$$M\begin{pmatrix}\alpha^1\\ \alpha^2\\ \alpha^3\end{pmatrix}=\begin{pmatrix}0, & -g\varphi^3-m_1, & g\varphi^2\\ -g\varphi^3-m_1 & 0, & -g\varphi^1\\ \partial_\mu A^2_\mu+A^2_\mu\partial_\mu, & -A^1_\mu\partial_\mu-\partial_\mu A^1_\mu, & \Box\end{pmatrix}\begin{pmatrix}\alpha_1\\ \alpha_2\\ \alpha_3\end{pmatrix}. \tag{3.21}$$

At small φ the determinant of the operator M is

$$\det M=m_1^2\det\Box+O(\varphi). \tag{3.22}$$

Since the first term is not zero, in the framework of perturbation theory $\det M\neq 0$, and the admissibility condition is fulfilled.

We shall now write down explicitly the quadratic form determining the mass spectrum

$$\mathscr{L}_0=-\frac{1}{4}(\partial_\nu A^a_\mu-\partial_\mu A^a_\nu)^2+\frac{m_1^2}{2}((A^1_\mu)^2+(A^2_\mu)^2)+$$
$$+\frac{1}{2}\partial_\mu\varphi^3\partial_\mu\varphi^3-\frac{m_2^2}{2}(\varphi^3)^2. \tag{3.23}$$

As is seen, our theory in the classical approximation describes two massive vector fields, one massless vector field, and one massive scalar particle. Therefore, indeed, two vector fields have acquired masses; however, quanta of two scalar fields have disappeared from the list of particles.

It is not difficult to construct an $SU(2)$-gauge-invariant model in which all three vector fields acquire a nonzero mass. For this it is necessary to examine the complex scalar field multiplet in the two-dimensional (spinor) representation

$$\varphi = \begin{pmatrix} \varphi_1 \\ \varphi_2 \end{pmatrix}, \quad \varphi^+ = (\varphi_1^*, \varphi_2^*). \tag{3.24}$$

The gauge-invariant Lagrangian has the form

$$\mathcal{L} = \mathcal{L}_{YM} + (\nabla_\mu \varphi)^+ \nabla_\mu \varphi - \lambda^2 (\varphi^+ \varphi - \mu^2)^2, \tag{3.25}$$

where

$$\nabla_\mu \varphi = \partial_\mu \varphi + \frac{i}{2} g \tau^a A_\mu^a \varphi, \tag{3.26}$$

and the gauge transformation of the fields φ is given by the formula

$$\delta\varphi(x) = \frac{1}{2i} g \tau^a \alpha^a (x) \varphi(x). \tag{3.27}$$

As in the previous case, a stable extremum corresponds to a constant φ such that

$$\varphi^+ \varphi = \mu^2. \tag{3.28}$$

We see that in this case the set of stable extrema forms a three-dimensional sphere S^3. In order to remove the degeneracy, we choose as a minimum

$$\varphi(x) = \begin{pmatrix} 0 \\ \mu \end{pmatrix}. \tag{3.29}$$

It can be verified that the condition

$$\varphi_1(x) = 0; \quad \operatorname{Im} \varphi_2(x) = 0 \tag{3.30}$$

is an admissible gauge. In this gauge there remains only one scalar field, $\operatorname{Re} \varphi_2(x) = (1/\sqrt{2})\sigma(x)$. Passing to fields with zero asymptotes at infinity,

$$\sigma(x) \to \sqrt{2}\,\mu + \sigma(x), \tag{3.31}$$

we obtain the Lagrangian

$$\mathscr{L} = -\frac{1}{4} F^a_{\mu\nu} F^a_{\mu\nu} + \frac{m_1^2}{2} A^a_\mu A^a_\mu + \frac{1}{2} \partial_\mu \sigma \partial_\mu \sigma - \frac{1}{2} m_2^2 \sigma^2 +$$

$$+ \frac{m_1 g}{2} \sigma A^a_\mu A^a_\mu + \frac{g^2}{8} \sigma^2 A^a_\mu A^a_\mu - \frac{g m_2^2}{4 m_1} \sigma^3 - \frac{g^2 m_2^2}{32 m_1^2} \sigma^4,$$

$$m_1 = \frac{\mu g}{\sqrt{2}}; \quad m_2 = 2\lambda\mu, \tag{3.32}$$

which describes the interaction of three massive vector fields and one massive scalar field.

The above-described mechanism will further be used for construction of unified gauge-invariant models of weak and electromagnetic interactions. We have finished the discussion of the classical Yang–Mills theory, and we shall now proceed to its quantization.

CHAPTER 2

QUANTUM THEORY IN TERMS OF PATH INTEGRALS

There exists several approaches to the quantization of a field theory. Most frequently, quantization is carried out by the operator method, in which operators satisfying canonical commutation relations correspond to classical field configurations. There is, however, another approach, in which the quantum dynamics is described by the sum over all field configurations, known as the path integral. Within this approach Feynman first formulated a self-consistent, manifestly relativistic-invariant perturbation theory for quantum electrodynamics. This formalism has turned out to be most convenient for the quantization of gauge fields, since the relativity principle is taken into account in the simplest way: Integration must be performed not over all field configurations, but only over gauge-equivalent classes.

In this chapter we deal with the general formalism of the path integral. In the next chapter we shall discuss applications of this formalism to gauge fields.

2.1. THE PATH INTEGRAL OVER PHASE SPACE

We shall start by demonstrating the main ideas of the path-integral method, as applied, for example, to nonrelativistic quantum mechanics. We begin with the case of a system with one degree of freedom.

Let p and q be the canonical momentum and coordinate of a particle ($-\infty < p < \infty$, $-\infty < q < \infty$). In the operator method of quantization, corresponding to p and q there are operators P, Q for which, most frequently, two realizations are used—the coordinate and

L. D. Faddeev and A. A. Slavnov, Gauge Fields: Introduction to Quantum Theory, ISBN 0-8053-9016-2.

Copyright © 1980 by The Benjamin/Cummings Publishing Company, Advanced Book Program. All rights reserved. No part of this publication may be reproduced, stored in a retrieval system, or transmitted, in any form or by any means, electronic, mechanical photocopying, recording, or otherwise, without the prior permission of the publisher.

the momentum ones. In the coordinate representation these operators and their eigenfunctions have the form

$$\begin{aligned} &Q = x; && P = \frac{1}{i}\frac{d}{dx}; \\ &|q\rangle = \delta(x-q); && |p\rangle = \left(\frac{1}{2\pi}\right)^{1/2} e^{ipx}; \\ &Q\,|q\rangle = q\,|q\rangle; && P\,|p\rangle = p\,|p\rangle. \end{aligned} \tag{1.1}$$

The transformation functions from the coordinate to the momentum representation and vice versa are given by the formulas

$$\langle p\,|\,q\rangle = \left(\frac{1}{2\pi}\right)^{1/2} e^{-ipq}; \quad \langle q\,|\,p\rangle = \left(\frac{1}{2\pi}\right)^{1/2} e^{ipq}. \tag{1.2}$$

The dynamics of the system is described with the help of the Hamiltonian function $h(p,q)$. In quantum mechanics there is a correspondence between this function and the Hamiltonian operator

$$H = h(P, Q), \tag{1.3}$$

Here a certain procedure for ordering noncommuting operator arguments P and Q is assumed. We do not here discuss the general problem of ordering, but we shall come back to it after introducing the concept of the path integral. The formal reasoning we use does not depend upon the concrete choice of the ordering procedure. For definiteness, we shall assume all the P operators to be placed to the left of all the Q operators.

With such a convention, the matrix element of the Hamiltonian H between the states $\langle p\,|$, $|\,q\rangle$ is explicitly expressed in terms of the classical Hamiltonian function

$$\langle p\,|\,H\,|\,q\rangle = \left(\frac{1}{2\pi}\right)^{1/2} e^{-ipq} h(p, q). \tag{1.4}$$

Our aim is to calculate the evolution operator

$$U(t'', t') = \exp\{-iH(t''-t')\}. \tag{1.5}$$

We shall calculate its matrix element

$$\begin{aligned} &\langle q''\,|\,U(t'', t')\,|\,q'\rangle = \\ &\qquad = \langle q''\,|\exp\{-iH(t''-t')\}\,|\,q'\rangle = \langle q'', t''\,|\,q', t'\rangle, \end{aligned} \tag{1.6}$$

which can be called the kernel of the operator U in the coordinate representation.

For small $t'' - t'$ this is easy to do taking into account the previous formula. Indeed, in this case

$$\exp\{-iH(t''-t')\} \cong 1 - iH(t''-t'), \tag{1.7}$$

and the matrix element $\langle p | U(t'', t') | q\rangle$ is approximately equal to

$$\langle p | U(t'', t') | q\rangle \cong \left(\frac{1}{2\pi}\right)^{1/2} e^{-ipq}(1 - ih(p, q)(t''-t')) \cong$$
$$\cong \left(\frac{1}{2\pi}\right)^{1/2} \exp\{-ipq - ih(p, q)(t''-t')\}. \tag{1.8}$$

The kernel of the operator $U(t'', t')$ in the coordinate representation is easily calculated by means of the transformation function

$$\langle q'' | U(t'', t') | q'\rangle = \int \langle q'' | p\rangle \langle p | U(t'', t') | q'\rangle \, dp =$$
$$= \frac{1}{2\pi} \int \exp\{ip(q''-q') - ih(p, q')(t''-t')\} \, dp. \tag{1.9}$$

For a finite interval $t'' - t'$ this formula is, of course, incorrect. In this case one may proceed as follows. We divide the interval $t'' - t'$ into N steps, considering

$$\Delta t = \frac{t''-t'}{N} \tag{1.10}$$

to be sufficiently small to use the previous formula for the operator $\exp\{-iH \; \Delta t\}$. The operator $U(t'', t')$ is expressed in terms of $\exp\{-iH \; \Delta t\}$ by the formula

$$U(t'', t') = (\exp\{-iH\Delta t\})^N. \tag{1.11}$$

Replacing each factor to the right by its kernel and integrating over all intermediate states, we obtain

$$\langle q'' | U(t'', t') | q' \rangle \cong \int \exp\{i[p_N(q_N - q_{N-1}) + \ldots$$
$$\ldots + p_1(q_1 - q_0)] - i[h(p_N, q_{N-1}) + \ldots + h(p_1, q_0)]\Delta t\} \times$$
$$\times \frac{dp_N}{2\pi} \frac{dp_{N-1}\, dq_{N-1}}{2\pi} \cdots \frac{dp_1\, dq_1}{2\pi}; \tag{1.12}$$

here $q_N = q''$, $q_0 = q'$.

Now we pass over to the limit $N \to \infty$, $\Delta t \to 0$. The number of integration variables also approaches infinity, and we can consider that in the limit we integrate over the values of the functions $p(t)$, $q(t)$ for all t in the interval $t' < t < t''$. The function $q(t)$ is subject to the condition

$$q(t') = q'; \quad q(t'') = q''. \tag{1.13}$$

The argument of the exponential function in this limit transforms into the integral

$$A_{t'}^{t''} = \int_{t'}^{t''} (p(t)\, \dot{q}(t) - h(p(t), q(t)))\, dt, \tag{1.14}$$

that is, into the classical action on the interval (t', t''). Thus, we obtain the main result: the matrix element of the evolution operator is found by integrating the Feynman functional $\exp\{iA_{t'}^{t''}\}$ over all trajectories $p(t)$, $q(t)$ in the phase space with fixed values q' and q'' at $t = t'$ and $t = t''$, respectively. The integration measure may be written formally as

$$\frac{dp''}{2\pi} \prod_t \frac{dp(t)\, dq(t)}{2\pi}, \tag{1.15}$$

that is, it is expressed in terms of the product of Liouville measures over all t. We have thus constructed the quantum-mechanical expression in terms of classical action and measure only.

The same final result

$$\langle q'', t'' | q', t' \rangle = \langle q'' | U(t'', t') | q' \rangle =$$
$$= \int \exp\left\{ i \int_{t'}^{t''} (p\dot{q} - h(p, q))\, dt \right\} \prod_t \frac{dp\, dq}{2\pi} \tag{1.16}$$

would be achieved also if we had used another ordering procedure for the operator factors. At first sight we have succeeded in unambiguously constructing quantum mechanics entirely in terms of classical objects which are canonical invariants. Actually, this cannot be true, since the whole group of canonical transformations of classical mechanics does not act in quantum mechanics. The solution of this apparent paradox is based on the fact that, really, we have not given the definition of the path integral in internal terms without assuming the limit process. In order to attach real meaning to the path integral, it is necessary to define the concrete way to calculate it, and this really is equivalent to choosing the ordering procedure. In field theory one such method (at present the only one) is given by perturbation theory. A strict definition of the path integral in this case will be given subsequently. Meanwhile, we shall deal with the path integral, considering it to be finite-dimensional. We hope that formal manipulations with the path integral, to be encountered below, will help the reader to develop a sufficiently clear intuitive notion of this object.

Feynman himself used a somewhat different form of the path integral, namely, the integral over the trajectories in the coordinate space. The Feynman formula is obtained if the Hamiltonian is quadratic in momenta:

$$h=\frac{p^2}{2m}+v(q). \tag{1.17}$$

Indeed, in this case the integration over the variables p can be carried out explicitly. In the integral

$$\int \exp\left\{ i \int_{t'}^{t''} \left(p\dot{q}-\frac{p^2}{2m}-v(q)\right) dt \right\} \prod_t \frac{dp\,dq}{2\pi} \tag{1.18}$$

a shift

$$p(t) \to p(t)+m\dot{q}. \tag{1.19}$$

must be performed. Then the integration over p and q becomes separated, and we get the answer

$$\langle q'', t'' \mid q', t' \rangle = \frac{1}{N} \int \exp\left\{ i \int_{t'}^{t''} \left(\frac{\dot{q}^2}{2m}-v(q)\right) dt \right\} \prod_t dq, \tag{1.20}$$

where

$$N^{-1} = \int \exp\left\{ -i \int_{t'}^{t''} \frac{p^2}{2m}\, dt \right\} \prod_t \frac{dp}{2\pi}. \tag{1.21}$$

The normalizing factor N is obviously independent of q' and q'' and is a function only of the time $t''-t'$. Usually this factor is included in the definition of measure. From the derivation presented above it is clear that the second form of the path integral is less general. It is correct only for Hamiltonians which are quadratic in momenta. However, for most problems that are interesting from a physical point of view the Hamiltonian possesses this property, and therefore for these problems the two forms are equivalent.

The case of a system with many degrees of freedom can be treated analogously. Using the vector notation

$$\begin{aligned} &p = (p_1, \ldots, p_n), \quad q = (q^1, \ldots, q^n), \\ &p\dot{q} = \sum_i p_i \dot{q}^i; \quad \frac{dp\,dq}{2\pi} = \prod_i \frac{dp_i\,dq^i}{2\pi}, \end{aligned} \tag{1.22}$$

we can retain the formulas (1.16), (1.20) in this case also.

From the viewpoint of the Hamiltonian dynamics the quantum field theory is a system with an infinite number of degrees of freedom. For example, in the case of a scalar neutral field described by the Lagrangian

$$\mathscr{L} = \frac{1}{2}\partial_\mu\varphi\partial_\mu\varphi - \frac{m^2}{2}\varphi^2 - V(\varphi), \tag{1.23}$$

the points in the phase space are pairs of functions $\varphi(\boldsymbol{x})$, $\pi(\boldsymbol{x})$ which form an infinite set of canonical variables. The argument $\boldsymbol{x}$ plays the role of the index number of these variables. The Poisson brackets are given by the relations.

$$\begin{aligned} &\{\varphi(\boldsymbol{x}), \varphi(\boldsymbol{y})\} = 0; \quad \{\pi(\boldsymbol{x}), \pi(\boldsymbol{y})\} = 0; \\ &\{\varphi(\boldsymbol{x}), \pi(\boldsymbol{y})\} = \delta^{(3)}(\boldsymbol{x} - \boldsymbol{y}). \end{aligned} \tag{1.24}$$

There exist many representations for the operators $\varphi(\boldsymbol{x})$ and $\pi(\boldsymbol{x})$ corresponding to $\varphi(\boldsymbol{x})$ and $\pi(\boldsymbol{x})$ after quantization. One representation, the coordinate one, is diagonal for $\varphi(\boldsymbol{x})$; the state vectors are functionals $\Phi(\varphi(\boldsymbol{x}))$ of $\varphi(\boldsymbol{x})$, and

$$\varphi(\boldsymbol{x})\Phi(\varphi) = \varphi(\boldsymbol{x})\Phi(\varphi); \quad \pi(\boldsymbol{x})\Phi(\varphi) = \frac{1}{i}\frac{\delta}{\delta\varphi(\boldsymbol{x})}\Phi(\varphi). \tag{1.25}$$

More often representations in the Fock space are used, and we shall mention them below. The Hamiltonian has the form

$$H(\pi, \varphi) = \int \left[\frac{\pi^2(\boldsymbol{x})}{2} + \frac{1}{2}\partial_k\varphi(\boldsymbol{x})\,\partial_k\varphi(\boldsymbol{x}) + \frac{m^2}{2}\varphi^2(\boldsymbol{x}) + V(\varphi)\right] d^3x. \quad (1.26)$$

It is easy to verify that the Hamiltonian equations of motion

$$\frac{d}{dt}\varphi(x) = \frac{\delta H}{\delta\pi(x)} = \pi(x);$$
$$\frac{d}{dt}\pi(x) = -\frac{\delta H}{\delta\varphi(x)} = \Delta\varphi - V'(\varphi) - m^2\varphi \quad (1.27)$$

do indeed coincide with the usual equation for the scalar field

$$\Box\varphi + m^2\varphi = -V'(\varphi). \quad (1.28)$$

The formulas obtained, which express the evolution operator in terms of the path integral, can be directly applied to this case also. In the coordinate representation we have

$$\langle\varphi''(\boldsymbol{x}), t'' \mid \varphi'(\boldsymbol{x}), t'\rangle = \langle\varphi''(\boldsymbol{x}) \mid \exp\{-iH(t''-t')\} \mid \varphi'(\boldsymbol{x})\rangle =$$
$$= \int \exp\left\{i\int_{t'}^{t''}\left[\pi(\boldsymbol{x}, t)\,\partial_t\varphi(\boldsymbol{x}, t) - \frac{\pi^2(\boldsymbol{x}, t)}{2} - \frac{1}{2}(\partial_k\varphi(\boldsymbol{x}, t))^2 - \right.\right.$$
$$\left.\left. - \frac{m^2\varphi^2(\boldsymbol{x}, t)}{2} - V(\varphi(\boldsymbol{x}, t))\right] d^3x\,dt\right\} \prod_{\boldsymbol{x}, t} \frac{d\pi(\boldsymbol{x}, t)\,d\varphi(\boldsymbol{x}, t)}{2\pi} =$$
$$= \frac{1}{N}\int \exp\left\{i\int \mathscr{L}(\varphi)\,d^4x\right\} \prod_x d\varphi(x)$$
$$t' < x_0 < t''; \quad \varphi(\boldsymbol{x}, t'') = \varphi''(\boldsymbol{x}); \quad \varphi(\boldsymbol{x}, t') = \varphi'(\boldsymbol{x}). \quad (1.29)$$

In the second formula we have used relativistic notation $x = (\boldsymbol{x}, t)$. The only thing in this formula which is not Lorentz invariant is the domain of integration over $t' \leq x_0 \leq t''$. Eventually we shall be interested in the evolution operator for an infinite time interval, since it is precisely this operator which is needed for construction of the scattering matrix, defined by the formula

$$S = \lim_{\substack{t''\to\infty \\ t'\to-\infty}} e^{iH_0t''} e^{-iH(t''-t')} e^{-iH_0t'}, \quad (1.30)$$

Here H_0 is the energy operator for free motion; it is obtained from H by omitting the interaction term $V(\varphi)$.

The representation we used before is inconvenient for calculating this limit, because the expression for the operator $\exp\{-iH_0 t\}$ in this representation is rather cumbersome. A more convenient one is the so-called holomorphic representation, in which the creation operators are diagonal. The next section is devoted to the discussion of this representation.

2.2. THE PATH INTEGRAL IN THE HOLOMORPHIC REPRESENTATION

We shall again begin with the case of one degree of freedom. Let us consider, as an example, a harmonic oscillator with the Hamiltonian

$$h(p, q) = \frac{p^2}{2} + \frac{\omega^2 q^2}{2}. \tag{2.1}$$

We introduce complex coordinates

$$a^* = \frac{1}{\sqrt{2\omega}}(\omega q - ip); \quad a = \frac{1}{\sqrt{2\omega}}(\omega q + ip); \tag{2.2}$$

in terms of these coordinates the Hamiltonian has the form $h = \omega a^* a$.

In quantum mechanics to these coordinates correspond operators, which are conjugates of each other and obey the commutation rules

$$[\boldsymbol{a}, \boldsymbol{a}^*] = 1. \tag{2.3}$$

These commutation relations have a representation in the space of analytic functions $f(a^*)$ with the scalar product

$$(f_1, f_2) = \int (f_1(a^*))^* f_2(a^*) e^{-a^* a} \frac{da^* \, da}{2\pi i}. \tag{2.4}$$

The operators $\boldsymbol{a}^*$ and $\boldsymbol{a}$ act in the following way:

$$\boldsymbol{a}^* f(a^*) = a^* f(a^*); \quad \boldsymbol{a} f(a^*) = \frac{d}{da^*} f(a^*). \tag{2.5}$$

Here we use the relation

$$\frac{da^* \, da}{2\pi i} = \frac{dp \, dq}{2\pi}. \tag{2.6}$$

The introduced scalar product is positive definite. Indeed, any arbitrary analytic function $f(a^*)$ is a linear combination of the monomials

$$\psi_n(a^*) = \frac{(a^*)^n}{\sqrt{n!}}. \tag{2.7}$$

A simple calculation shows that these monomials are orthonormalized:

$$\langle\psi_n | \psi_m\rangle = \frac{1}{\sqrt{n!m!}} \int a^n (a^*)^m e^{-a^*a} \frac{da^*\, da}{2\pi i} =$$

$$= \frac{1}{\sqrt{m!n!}} \frac{1}{\pi} \int_0^\infty \rho\, d\rho \int_0^{2\pi} d\theta \rho^{n+m} e^{i\theta(n-m)} e^{-\rho^2} = \begin{cases} 0, & n \neq m, \\ 1, & n = m, \end{cases} \tag{2.8}$$

whence there follows the positive definiteness of the scalar product.

It is also clear that the operators $\boldsymbol{a}^*$ and $\boldsymbol{a}$ are conjugates of each other. Indeed, taking into account that

$$a^* e^{-a^*a} = -\frac{d}{da} e^{-a^*a}; \quad \frac{d}{da} f(a^*) = 0, \tag{2.9}$$

and integrating by parts, we have

$$(f_1, \boldsymbol{a}^* f_2) = \int (f_1(a^*))^* a^* f_2(a^*) e^{-a^*a} \frac{da^*\, da}{2\pi i} =$$

$$= \int \left[\frac{d}{da} ((f_1(a^*))^* f_2(a^*))\right] e^{-a^*a} \frac{da^*\, da}{2\pi i} =$$

$$= \int \left(\frac{d}{da^*} f_1(a^*)\right)^* f_2(a^*) e^{-a^*a} \frac{da^*\, da}{2\pi i} = (\boldsymbol{a} f_1, f_2). \tag{2.10}$$

There are two ways to describe arbitrary operators in this representation. First, an arbitrary operator $\boldsymbol{A}$ can be represented by an integral operator with a kernel $A(a^*, a)$

$$(\boldsymbol{A} f)(a^*) = \int A(a^*, \alpha) f(\alpha^*) e^{-\alpha^*\alpha} \frac{d\alpha^*\, d\alpha}{2\pi i}. \tag{2.11}$$

The kernel $A(a^*, a)$ is expressed in terms of the matrix elements of the operator A in the basis ψ_n: if

$$A_{nm} = \langle\psi_n | \boldsymbol{A} | \psi_m\rangle, \tag{2.12}$$

then

$$A(a^*, a) = \sum_{n, m} A_{nm} \frac{(a^*)^n}{\sqrt{n!}} \frac{a^m}{\sqrt{m!}}. \tag{2.13}$$

This formula defines $A(a^*, a)$ as an analytic function of two complex variables, a^*, a which are not necessary conjugates of each other.

The convolution of the kernels corresponds to the product of the operators $\boldsymbol{A}_1$ and $\boldsymbol{A}_2$:

$$(A_1 A_2)(a^*, a) = \int A_1(a^*, \alpha) A_2(\alpha^*, a) e^{-\alpha^* \alpha} \frac{d\alpha^* d\alpha}{2\pi i}. \tag{2.14}$$

The second representation for operators is simply the definition of an operator in the form of a normally ordered polynomial in operators $\boldsymbol{a}^*$ and $\boldsymbol{a}$. A product in which all operators $\boldsymbol{a}^*$ are placed to the left of all operators $\boldsymbol{a}$ is called a normal product. Let us examine the kernel of an operator $\boldsymbol{A}$, given in terms of a sum of normal products:

$$\boldsymbol{A} = \sum_{n, m} K_{nm} (\boldsymbol{a}^*)^n \boldsymbol{a}^m. \tag{2.15}$$

This operator can be associated with a function

$$K(a^*, a) = \sum_{n, m} K_{nm} (a^*)^n a^m, \tag{2.16}$$

We shall call this function the normal symbol of the operator $\boldsymbol{A}$. Then the kernel $A(a^*, a)$ of the operator $\boldsymbol{A}$ is related to $K(a^*, a)$ by the formula

$$A(a^*, a) = e^{a^* a} K(a^*, a). \tag{2.17}$$

For a check we shall consider, as an operator $\boldsymbol{A}$, the monomial

$$\boldsymbol{A} = (\boldsymbol{a}^*)^k \boldsymbol{a}^l, \tag{2.18}$$

so that

$$K(a^*, a) = (a^*)^k a^l \tag{2.19}$$

and

$$A_{nm} = \langle \psi_n | \boldsymbol{A} | \psi_m \rangle =$$
$$= \frac{1}{\sqrt{n!m!}} \int \left(\left(\frac{d}{da^*} \right)^k (a^*)^n \right)^* \left(\left(\frac{d}{da^*} \right)^l (a^*)^m \right) e^{-a^*a} \frac{da^*\, da}{2\pi i} =$$
$$= \sqrt{n(n-1)\ldots(n-k+1)}\, \sqrt{m(m-1)\ldots(m-l+1)} \times$$
$$\times \theta(n \geqslant k)\, \theta(m \geqslant l)\, \delta_{n-k,\, m-l}, \qquad (2.20)$$

where

$$\theta(n \geqslant k) = \begin{cases} 0, & \text{if} \quad n < k, \\ 1, & \text{if} \quad n \geqslant k. \end{cases}$$

We shall now construct $A(a^*, a)$, using the formula (2.13). We have

$$A(a^*, a) = \sum_{n,\, m} A_{nm} \frac{(a^*)^n}{\sqrt{n!}} \frac{a^m}{\sqrt{m!}} = (a^*)^k a^l \sum \frac{(a^*)^n a^n}{n!} =$$
$$= (a^*)^k a^l \exp\{a^* a\}. \qquad (2.21)$$

The formula (2.17) is thus verified. The formulas (2.17) and (2.14) allow us in a simple way to construct the evolution operator in the form of a path integral over the functions $a^*(t)$ and $a(t)$. The corresponding derivation is actually a repetition of the reasoning of Section 2.1.

Let the Hamiltonian be given in the form

$$H = h(\boldsymbol{a}^*, \boldsymbol{a}), \qquad (2.22)$$

where a normal ordering is assumed. Then the kernel $U(a^*, a, \Delta t)$ of the evolution operator

$$U(\Delta t) = \exp\{-iH\Delta t\} \qquad (2.23)$$

for small Δt takes the following form:

$$U(a^*, a, \Delta t) = \exp\{a^* a - ih(a^*, a)\Delta t\}. \qquad (2.24)$$

For an arbitrary interval $t'' - t' = N\Delta t$ we must calculate the convolution of N such kernels:

$$U(a^*, a; t''-t') =$$
$$= \int \exp\{[a_N^* a_{N-1} - a_{N-1}^* a_{N-1} + \ldots - a_1^* a_1 + a_1^* a_0] -$$
$$- i[h(a_N^*, a_{N-1}) + \ldots + h(a_1^*, a_0)]\Delta t\} \prod_{k=1}^{N-1} \frac{da_k^* \, da_k}{2\pi i}, \tag{2.25}$$

where we have denoted $a_0 = a$, $a_N = a^*$. The formal limit as $\Delta t \to 0$ and $N \to \infty$ is expressed by

$$U(a^*, a; t''-t') = \int \exp\{a^*(t'')\, a(t'')\} \times$$
$$\times \exp\left\{\int_{t'}^{t''} (-a^*\dot{a} - ih(a^*, a))\, dt\right\} \prod_t \frac{da^*\, da}{2\pi i}, \tag{2.26}$$

or, after symmetrizing in a^* and a,

$$U(a^*, a; t''-t') =$$
$$= \int \exp\left\{\frac{1}{2}(a^*(t'')\, a(t'') + a^*(t')\, a(t'))\right\} \times$$
$$\times \exp\left\{ i \int_{t'}^{t''} \left[\frac{1}{2i}(\dot{a}^* a - a^* \dot{a}) - h(a^*, a)\right] dt \right\} \prod_t \frac{da^*\, da}{2\pi i}. \tag{2.27}$$

Here it is assumed that $a^*(t'') = a^*$ and $a(t') = a$. We may point out that the latter formula differs insignificantly from the corresponding formula (1.16) of the previous section. In both these formulas the integrand is the functional $\exp\{i \times \text{action}\}$, and the integration is carried over the product of Liouville measures over the phase space. The additional functional

$$\exp\left\{\frac{1}{2}(a^*(t'')\, a(t'') + a^*(t')\, a(t'))\right\} \tag{2.28}$$

in the formula (2.27) reflects the differences in boundary conditions on the trajectories over which we integrate: in the case of (1.16) we fix the value of the same function $q(t)$ at $t = t'$ and $t = t''$, whereas in the case of (2.27) at $t = t'$ the value of the function $a(t)$ is fixed, and $t = t''$ it is the value of the function $a^*(t)$. We emphasize that the variables $a^*(t'')$ and $a(t'')$ are independent; we integrate over $a(t'')$, but $a^*(t'')$ remains fixed. Analogously, we integrate over the variable $a^*(t')$, leaving $a(t')$ fixed.

In the case of a harmonic oscillator the integral (2.27) is easily computed, since the integrand is an exponential function of a non-uniform quadratic form. We shall call such integrals Gaussian integrals. We shall take advantage of the well-known property of a Gaussian integral, according to which it is equal to the value of the integrand, calculated at the extremum point of the power of the exponential function. The condition for the extremum in our case coincides with the classical equation of motion

$$\dot{a}^* - i\omega a^* = 0; \quad \dot{a} + i\omega a = 0; \quad a^*(t'') = a^*; \quad a(t') = a, \tag{2.29}$$

since

$$\delta\left(a^*(t'')\, a(t'') + \int_{t'}^{t''} (-a^*\dot{a} - i\omega\, a^* a)\, dt\right) = \int_{t'}^{t''} (\delta a\, (\dot{a}^* - i\omega a^*) - \delta a^*\, (\dot{a} + i\omega a))\, dt \tag{2.30}$$

at $\delta a^*\big|_{t''} = 0$; $\delta a\big|_{t'} = 0$.

The equations (2.29) can be solved in a trivial manner:

$$a(t) = e^{i\omega(t'-t)} a; \quad a^*(t) = e^{i\omega(t-t'')} a^*. \tag{2.31}$$

Denoting the corresponding evolution operator by U_0, we obtain

$$U_0(a^*,\ a;\ t'' - t') = \exp\{a^* a\, e^{i\omega(t'-t'')}\}. \tag{2.32}$$

if $f(a^*)$ is an arbitrary function, then

$$U_0(t)\, f(a^*) = \int \exp\{a^* \alpha\, e^{-i\omega t}\}\, f(\alpha^*)\, e^{-\alpha^* \alpha}\, \frac{d\alpha^*\, d\alpha}{2\pi i} = f(a^* e^{-i\omega t}). \tag{2.33}$$

This formula clearly demonstrates the convenience of using the holomorphic representation for a harmonic oscillator. In this representation the evolution of an arbitrary state is reduced to the substitution of the argument

$$a^* \rightarrow a^* e^{-i\omega t}. \tag{2.34}$$

This property is very useful for the field theory, because in this case the free Hamiltonian is represented by the sum of the Hamiltonians of an infinite set of oscillators.

2.3. THE GENERATING FUNCTIONAL FOR THE S-MATRIX IN FIELD THEORY

The holomorphic representation is introduced in the field theory in terms of complex amplitudes $a^*(\boldsymbol{k})$ and $a(\boldsymbol{k})$. The canonical variables $\varphi(\boldsymbol{x})$ and $\pi(\boldsymbol{x})$ are expressed in terms of these amplitudes in the following way:

$$\begin{gathered}
\varphi(\boldsymbol{x}) = \left(\frac{1}{2\pi}\right)^{3/2} \int (a^*(\boldsymbol{k}) e^{-i\boldsymbol{k}\boldsymbol{x}} + a(\boldsymbol{k}) e^{i\boldsymbol{k}\boldsymbol{x}}) \frac{d^3k}{\sqrt{2\omega}}, \\
\pi(\boldsymbol{x}) = \left(\frac{1}{2\pi}\right)^{3/2} \int (a^*(\boldsymbol{k}) e^{-i\boldsymbol{k}\boldsymbol{x}} - a(\boldsymbol{k}) e^{i\boldsymbol{k}\boldsymbol{x}})\, i \sqrt{\frac{\omega}{2}}\, d^3k, \\
k_0 = \omega = (k^2 + m^2)^{1/2}.
\end{gathered} \tag{3.1}$$

Under quantization the amplitudes $a^*(\boldsymbol{k})$ and $a(\boldsymbol{k})$ acquire the meaning of creation and annihilation operators, respectively.

The free Hamiltonian H_0 is expressed in terms of $\boldsymbol{a}^*$, $\boldsymbol{a}$ as follows:

$$H_0 = \int \omega(k)\, \boldsymbol{a}^*(\boldsymbol{k})\, \boldsymbol{a}(\boldsymbol{k})\, d^3k \tag{3.2}$$

and is the sum of the energies of an infinite set of oscillators. The argument $\boldsymbol{k}$ plays the role of the oscillator index number and $\omega(\boldsymbol{k})$ of its frequency. The complete Hamiltonian H, besides the term H_0, contains the interaction $V(\boldsymbol{a}^*, \boldsymbol{a})$, which is obtained by substitution of the function $\varphi(\boldsymbol{x})$ in the form (3.1) into $\int V(\varphi) d^3x$. The evolution operator $U(t'', t')$ is determined by the kernel $U(a^*(\boldsymbol{k}), a(\boldsymbol{k}), t'' - t')$, which is expressed in terms of the path integral

$$U(a^*(\boldsymbol{k}),\ a(\boldsymbol{k}),\ t''-t')=$$
$$=\int \exp\Big\{\int d^3k a^*(\boldsymbol{k},\ t'')\, a(\boldsymbol{k},\ t'')+\int_{t'}^{t''}[-iV(a^*,\ a)+$$
$$+\int d^3k\,(-a^*(\boldsymbol{k},\ t)\,\dot{a}(\boldsymbol{k},\ t)-i\omega a^*(\boldsymbol{k},\ t)\, a(\boldsymbol{k},\ t))\Big]\,dt\Big\}\times$$
$$\times\prod_{t,\ k}\frac{da^*(\boldsymbol{k},\ t)\, da(\boldsymbol{k},\ t)}{2\pi i};\quad a^*(\boldsymbol{k},\ t'')=a^*(\boldsymbol{k});\ a(\boldsymbol{k},\ t')=a(\boldsymbol{k}).\tag{3.3}$$

From this formula it is easy to pass over to the S-matrix. For this we point out that for an arbitrary operator A with a kernel $A(a^*(\boldsymbol{k}),\ a(\boldsymbol{k}))$ the operator

$$e^{iH_0t''}Ae^{-iH_0t'}\tag{3.4}$$

has the kernel

$$A\,(a^*(\boldsymbol{k})\,e^{i\omega t''},\ a(\boldsymbol{k})\,e^{-i\omega t'}).\tag{3.5}$$

This is a direct generalization of the formula (2.33) for a harmonic oscillator, obtained in the previous section. Thus, the kernel of the S-matrix is obtained as the limit for $t''\to\infty$ and $t'\to-\infty$ of the integral (3.3), which we shall for convenience rewrite in a symmetrized form as

$$S\,(a^*(k),\ a(k))=$$
$$=\lim_{\substack{t''\to\infty\\ t'\to-\infty}}\int \exp\Big\{\frac{1}{2}\int d^3k\,(a^*(\boldsymbol{k},\ t'')\, a(\boldsymbol{k},\ t'')+a^*(\boldsymbol{k},\ t')\, a(\boldsymbol{k},\ t'))+$$
$$+i\int_{t'}^{t''}dt\Big[\Big[\int d^3\boldsymbol{k}\Big(\frac{1}{2i}(\dot{a}^*(\boldsymbol{k},\ t)\, a(\boldsymbol{k},\ t)-a^*(\boldsymbol{k},\ t)\,\dot{a}(\boldsymbol{k},\ t))-$$
$$-\omega(\boldsymbol{k})\, a^*(\boldsymbol{k},\ t)\, a(\boldsymbol{k},\ t)\Big)-V(a^*,\ a)]]\Big\}\prod_{k,\ t}\frac{da^*(\boldsymbol{k},\ t)\, da(\boldsymbol{k},\ t)}{2\pi i},\tag{3.6}$$

where

$$a^*(\boldsymbol{k},\ t'')=a^*(\boldsymbol{k})\exp\{i\omega(\boldsymbol{k})\, t''\},\tag{3.7}$$

$$a(\boldsymbol{k}, t') = a(\boldsymbol{k}) \exp\{-i\omega(\boldsymbol{k}) t'\}. \tag{3.8}$$

We shall apply this formula to calculating the S-matrix for scattering on an external source $\eta(x)$, when

$$V(\varphi) = -\eta(x)\varphi(x). \tag{3.9}$$

The corresponding functional $V(a^*, a)$ has the form

$$V(a^*, a) = \int d^3k\, [\gamma(\boldsymbol{k}, t)\, a^*(\boldsymbol{k}) + \gamma^*(\boldsymbol{k}, t)\, a(\boldsymbol{k})], \tag{3.10}$$

where

$$\gamma(\boldsymbol{k}, t) = -\frac{1}{\sqrt{2k_0}}\left(\frac{1}{2\pi}\right)^{3/2} \int \eta(\boldsymbol{x}, t)\, e^{-i\boldsymbol{k}\boldsymbol{x}}\, d^3x. \tag{3.11}$$

The functional $V(a^*, a)$ depends explicitly on time. Nevertheless, all the formulas for the evolution operator in this case still remain valid. The only change is that the evolution operator now depends on both variables t'' and t' and not only on the difference between them. The integrand in (3.6) in our case again takes the form of an exponential function of a nonuniform quadratic form, and the Gaussian integral is calculated in the same manner as in the previous section. The conditions for the extremum are the following:

$$\begin{gathered}
\dot{a}(\boldsymbol{k}, t) + i\omega(\boldsymbol{k})\, a(\boldsymbol{k}, t) + i\gamma(\boldsymbol{k}, t) = 0, \\
\dot{a}^*(\boldsymbol{k}, t) - i\omega(\boldsymbol{k})\, a^*(\boldsymbol{k}, t) - i\gamma^*(\boldsymbol{k}, t) = 0, \\
a^*(\boldsymbol{k}, t'') = a^*(\boldsymbol{k})\, e^{i\omega t''}, \quad a(\boldsymbol{k}, t') = a(\boldsymbol{k})\, e^{-i\omega t'}.
\end{gathered} \tag{3.12}$$

The solution of these equations is given by the formulas

$$a^*(\boldsymbol{k}, t) = a^*(\boldsymbol{k})\, e^{i\omega t} - i e^{i\omega t} \int_t^{t''} e^{-i\omega s} \gamma^*(\boldsymbol{k}, s)\, ds, \tag{3.13}$$

$$a(\boldsymbol{k}, t) = a(\boldsymbol{k})\, e^{-i\omega t} - i e^{-i\omega t} \int_{t'}^{t} e^{i\omega s} \gamma(\boldsymbol{k}, s)\, ds. \tag{3.14}$$

Substituting this solution into the exponential function in the formula (3.6) and passing to the limit, we obtain the following expression for the kernel of the S-matrix:

$$S_\eta(a^*, a) = \exp\Big\{ \int d^3k \Big[a^*(\boldsymbol{k})\, a(\boldsymbol{k}) +$$

$$+ \frac{1}{(2\pi)^{3/2}} \int_{-\infty}^{\infty} dt \int d^3x\, \eta(\boldsymbol{x}, t)\, \frac{a^*(\boldsymbol{k}) e^{i\omega t} e^{-i(\boldsymbol{k}\boldsymbol{x})} + a(\boldsymbol{k})\, e^{-i\omega t} e^{i(\boldsymbol{k}\boldsymbol{x})}}{\sqrt{2\omega}} -$$

$$- \left(\frac{1}{2\pi}\right)^3 \frac{1}{2} \int_{-\infty}^{\infty} dt \int_{-\infty}^{\infty} ds \int d^3x \int d^3y \frac{1}{2\omega}\, \eta(\boldsymbol{x}, t) \times$$

$$\times\, \eta(\boldsymbol{y}, s) e^{i\boldsymbol{k}(\boldsymbol{x}-\boldsymbol{y})} e^{-i\omega|t-s|} \Big] \Big\}. \tag{3.15}$$

The expression for the S-matrix becomes more elegant if we pass from the kernel to the normal symbol, which is equivalent to omitting the first factor $\exp\{\int d^3k \cdot a^*(\boldsymbol{k}) a(\boldsymbol{k})\}$. The remaining factors may be rewritten in a manifestly relativistic-invariant form. For this we introduce the solution of the free Klein–Gordon equation

$$\varphi_0(x) = \left(\frac{1}{2\pi}\right)^{3/2} \int (a^*(\boldsymbol{k})\, e^{ikx} + a(\boldsymbol{k})\, e^{-ikx}) \frac{d^3k}{\sqrt{2k_0}}, \quad k_0 = \omega \tag{3.16}$$

$$\Box\, \varphi_0 + m^2 \varphi_0 = 0 \tag{3.17}$$

and the Green function of this equation

$$D_c(x) = -\left(\frac{1}{2\pi}\right)^3 \int e^{ik x} e^{-i\omega|x_0|} \frac{d^3k}{2i\omega} =$$

$$= -\left(\frac{1}{2\pi}\right)^4 \int e^{-ikx} \frac{1}{k^2 - m^2 + i0}\, d^4k, \tag{3.18}$$

$$(\Box + m^2)\, D_c = \delta^4(x). \tag{3.19}$$

The first representation for D_c follows from the second one upon integration over k_0.

The normal symbol of the S-matrix $S_\eta(a^*,a)$ in the above notation is given by the formula

$$S_\eta(a^*, a) =$$

$$= \exp\Big\{ i \int \eta(x)\, \varphi_0(x)\, dx + \frac{i}{2} \int \eta(x)\, D_c(x-y)\, \eta(y)\, dx\, dy \Big\}. \tag{3.20}$$

We may point out that the proper choice of asymptotic conditions on the trajectory of integration has led to the appearance of the causal Green function in the formula for the S-matrix.

Now let us pass to the consideration of the S-matrix in the case of the general potential $V(\varphi)$. In this case we evidently cannot calculate the corresponding path integral exactly, and we shall restrict ourselves to the construction of a perturbation theory for it. We shall show that in this case the problem is reduced to the already solved problem concerning the scattering on an external field. For this purpose we shall make use of the obvious formula

$$\varphi(x_1)\dots\varphi(x_n) = \\ = \frac{1}{i}\frac{\delta}{\delta\eta(x_1)}\dots\frac{1}{i}\frac{\delta}{\delta\eta(x_n)}\exp\left\{i\int\varphi(x)\eta(x)\,dx\right\}\Big|_{\eta=0}; \quad (3.21)$$

From this formula it follows that an arbitrary functional $\Phi(\varphi)$ of $\varphi(x)$ can be written in the form

$$\Phi(\varphi) = \Phi\left(\frac{1}{i}\frac{\delta}{\delta\eta(x)}\right)\exp\left\{i\int\varphi(x)\eta(x)\,dx\right\}\Big|_{\eta=0}. \quad (3.22)$$

In particular,

$$\exp\left\{-i\int V(\varphi)\,dx\right\} = \\ = \exp\left\{-i\int V\left(\frac{1}{i}\frac{\delta}{\delta\eta}\right)dx\right\}\exp\left\{i\int\varphi\eta\,dx\right\}\Big|_{\eta=0}. \quad (3.23)$$

This formula, of course, is understood in the sense of perturbation theory.

Thus, in the path integral (3.6), which determines the S-matrix for a potential of the general form, we can substitute the $\exp\{-i\int V(\varphi)\,dx\}$ in the integrand by the right-hand side of (3.23) and put outside of the path integral the formal differential operator

$$\exp\left\{-i\int V\left(\frac{1}{i}\frac{\delta}{\delta\eta}\right)dx\right\}$$

The remaining path integral coincides precisely with the already calculated integral for the S-matrix for scattering on an external source. As a result, we obtain the following final expression for the normal symbol of the S-matrix:

$$S(a^*, a) = S(\varphi_0) =$$
$$= \exp\left\{-i \int V\left(\frac{1}{i}\frac{\delta}{\delta\eta(x)}\right) dx\right\} \exp\left\{i \int \eta(x)\,\varphi_0(x)\,dx + \right.$$
$$\left. + \frac{i}{2}\int \eta(x)\,D_c(x-y)\,\eta(y)\,dx\,dy\right\}\Bigg|_{\eta=0}. \quad (3.24)$$

Here we have replaced the pair of arguments a^*, a by a single function φ_0, since they define each other uniquely. Expanding this functional in a series in φ_0,

$$S(\varphi_0) =$$
$$= \sum_n \frac{1}{n!}\int S_n(x_1 \ldots x_n)\,\varphi_0(x_1)\ldots\varphi_0(x_n)\,dx_1\ldots dx_n \quad (3.25)$$

we obtain the coefficient functions $S_n(x_1,\ldots,x_n)$. In the operator formalism these functions appear when the S-matrix operator is expanded in a series over the normal products of free fields. For this reason the functional $S(\varphi_0)$ is sometimes called the generating functional for the coefficient functions of the S-matrix.

We leave it to the reader to verify that the expansion (3.24) in a perturbation-theory series generates usual Feynman diagrams. The function $D_c(x-y)$ plays the role of the propagator, the vertices are defined by the potential $V(\varphi)$, and the function φ_0 corresponds to the external lines. Thus, the formula (3.24) automatically takes into account Wick's theorem for chronological products.

If in the formula (3.24) we do not assume $\eta = 0$, then the resulting functional

$$S(\varphi_0, \eta) =$$
$$= \exp\left\{-i \int V\left(\frac{1}{i}\frac{\delta}{\delta\eta(x)}\right) dx\, \exp\left\{i \int \eta(x)\,\varphi_0(x)\,dx\right\}\times\right.$$
$$\times \exp\left\{\frac{i}{2}\int \eta(x)\,D_c(x-y)\,\eta(y)\,dx\,dy\right\} \quad (3.26)$$

is the normal symbol of the S-matrix for the scattering of interacting particles in the presence of an external source $\eta(x)$. In practice, it is often more convenient to deal not with the S-matrix (3.24), but with the functional

$$Z(\eta) = \exp\left\{-i\int V\left(\frac{1}{i}\frac{\delta}{\delta\eta(x)}\right)dx\right\}\times$$
$$\times\exp\left\{\frac{i}{2}\int\eta(x)D_c(x-y)\eta(y)\,dx\,dy\right\}, \quad (3.27)$$

which coincides with $S(\varphi_0, \eta)$ at $\varphi_0 = 0$ and has the meaning of the transition amplitude from vacuum to vacuum in the presence of an external source. The coefficient functions $G_n(x_1,\ldots,x_n)$ in the expansion of this functional in the series in $\eta(x)$,

$$Z(\eta) = \sum\frac{1}{n!}\int G_n(x_1, \ldots, x_n)\,\eta(x_1)\ldots\eta(x_n)\,dx_1\ldots dx_n \quad (3.28)$$

define the so-called Green functions, to which in the operator formalism the average values of the chronological products of the Heisenberg field operators correspond. The Green functions are necessary, in particular, for realization of the renormalization program, which will be discussed in the next chapters and which until now has not yet been formulated for the S-matrix directly.

The functional $Z(\eta)$ itself contains more information than $S(\varphi_0)$, since it is defined for arbitrary functions η, whereas $S(\varphi_0)$ is defined only on the mass shell, that is, its argument φ_0 is the solution of the free-field equation of motion. Knowledge of the functional $Z(\eta)$ allows the reconstruction of $S(\varphi_0)$. The corresponding procedure is determined by the so-called reduction formulas, which are readily deduced from comparison of the formulas (3.24) and (3.27).

In order to obtain explicit formulas, we introduce the extended functional $\tilde{S}(\varphi)$ by replacing φ_0 in (3.24) with an arbitrary function of four variables. Then the coefficient functions extended off the mass shell are variational derivatives

$$\tilde{S}_n(x_1, \ldots, x_n) = \frac{1}{i}\frac{\delta}{\delta\varphi(x_1)}\cdots\frac{1}{i}\frac{\delta}{\delta\varphi(x_n)}\tilde{S}(\varphi)\Big|_{\varphi=0}. \quad (3.29)$$

On the other hand, we can replace the argument $\eta(x)$ in the functional $Z(\eta)$ by $\tilde{\eta}(x)$, where

$$\tilde{\eta}(x) = \int D_c(x-y)\,\eta(y)\,dy. \quad (3.30)$$

Then by direct comparison we verify that

$$\int \prod_i dx_i \varphi_0(x_i) \left\{ \frac{1}{i} \frac{\delta}{\delta \varphi(x_1)} \cdots \frac{1}{i} \frac{\delta}{\delta \varphi(x_n)} \tilde{S}(\varphi) \Big|_{\varphi=0} - \right.$$

$$\left. - \frac{1}{i} \frac{\delta}{\delta \tilde{\eta}(x_1)} \cdots \frac{1}{i} \frac{\delta}{\delta \tilde{\eta}(x_n)} Z(\tilde{\eta}) \Big|_{\eta=0} \right\} = 0. \quad (3.31)$$

Thus, we have a simple procedure for calculating the normal symbol of the S-matrix. One must calculate the variational derivatives

$$\frac{1}{i} \frac{\delta}{\delta \eta(x_1)} \cdots \frac{1}{i} \frac{\delta}{\delta \eta(x_n)} Z(\eta) \Big|_{\eta=0}, \quad (3.32)$$

(that is, the Green functions $G_n(x_1, \ldots, x_n)$), apply to these functions the differential operator

$$\prod_{i=1}^{n} (\Box_{x_i} + m^2), \quad (3.33)$$

then multiply the result by the product

$$\frac{1}{n!} \prod_i \varphi_0(x_i), \quad (3.34)$$

integrate over all x, and sum over n.

An alternative formalism for calculating the S-matrix may be based directly on its representation in the form of a path integral. The expression (3.6) is inconvenient for this purpose, since it is not manifestly relativistic-invariant and it involves a limit process. Let us transform this expression into a manifestly relativistic form, by integrating over the momenta $\pi(\mathbf{x}, t)$. It is necessary, however, to take into account accurately the boundary terms. We point out, first of all, that the action functional in the formula (3.6) can be rewritten in terms of the fields $\varphi(\mathbf{x}, t)$ and $\pi(\mathbf{x}, t)$

$$\int d^3x \int_{t'}^{t''} \left[\frac{1}{2} (\pi \, \partial_0 \varphi - \partial_0 \pi \varphi) - \right.$$

$$\left. - \frac{1}{2} \pi^2 - \frac{1}{2} (\partial_k \varphi)^2 - \frac{1}{2} m^2 \varphi^2 - V(\varphi) \right] dt, \quad (3.35)$$

and the integration measure in terms of φ and π has the form

$$\prod_{\substack{\mathbf{k},\, t \\ t' \leqslant t \leqslant t''}} \frac{da^*(\mathbf{k}, t)\, da(\mathbf{k}, t)}{2\pi i} = \prod_{\substack{\mathbf{x},\, t \\ t' \leqslant t \leqslant t''}} \frac{d\varphi(\mathbf{x}, t)\, d\pi(\mathbf{x}, t)}{2\pi}. \quad (3.36)$$

Here we have used the relationship between the integration variables $a^*(\boldsymbol{k}, t)$, $a(\boldsymbol{k}, t)$ and $\pi(\boldsymbol{x}, t)$, $\varphi(\boldsymbol{x}, t)$:

$$\begin{aligned}
&\varphi(\boldsymbol{x}, t) = \\
&= \left(\frac{1}{2\pi}\right)^{3/2} \int (a^*(\boldsymbol{k}, t)\, e^{-ikx} + a(\boldsymbol{k}, t)\, e^{ikx}) \frac{d^3k}{\sqrt{2\omega}}, \qquad (3.37)\\
&\pi(\boldsymbol{x}, t) = \\
&= \left(\frac{1}{2\pi}\right)^{3/2} \int (a^*(\boldsymbol{k}, t)\, e^{-ikx} - a(\boldsymbol{k}, t)\, e^{ikx})\, i \sqrt{\frac{\omega}{2}}\, d^3k,
\end{aligned}$$

which has already been introduced at $t = 0$ in the formulas (3.1).

Let us use in the integral (3.6), together with the variables $\varphi(\boldsymbol{x}, t)$ and $\pi(\boldsymbol{x}, t)$, also the variables $\phi_1(\boldsymbol{x}, t)$ and $\pi_1(\boldsymbol{x}, t)$ which are obtained from the former ones by the shift

$$\begin{aligned}
\varphi(\boldsymbol{x}, t) &= \varphi_0(\boldsymbol{x}, t) + \varphi_1(\boldsymbol{x}, t),\\
\pi(\boldsymbol{x}, t) &= \partial_0\varphi(\boldsymbol{x}, t) + \pi_1(\boldsymbol{x}, t).
\end{aligned} \qquad (3.38)$$

Here $\varphi_0(\boldsymbol{x}, t) = \varphi_0(x)$ is constructed according to the formula (3.16) in terms of the functions $a^*(\boldsymbol{k})$ and $a(\boldsymbol{k})$ which enter into the boundary conditions (3.7) and (3.8). Integration by parts transforms the action (3.5) into the following form in terms of the new variables:

$$\begin{aligned}
&\int d^3x \left[\partial_0\varphi_0\varphi - \frac{1}{2}\partial_0\varphi_0\varphi_0 - \frac{1}{2}\pi\varphi\right]\Big|_{t'}^{t''} + \\
&\quad + \int d^3x \int_{t'}^{t''} dt \Big[-\frac{1}{2}\pi_1^2 + \frac{1}{2}(\partial_0\varphi_1\,\partial_0\varphi_1 - \partial_k\varphi_1\,\partial_k\varphi_1) - \\
&\qquad\qquad - \frac{1}{2}m^2\varphi_1^2 - V(\varphi)\Big]. \qquad (3.39)
\end{aligned}$$

We see, in particular, that in the second term the variables φ and π_1 are completely separated.

Using (3.37) and the definition (3.16) of the function $\varphi_0(x)$ and the boundary conditions (3.7), (3.8), we can rewrite the terms outside of the integrals in (3.39) as follows:

$$\begin{aligned}
&i\int d^3x \left[\partial_0\varphi_0\varphi - \frac{1}{2}\partial_0\varphi_0\varphi_0 - \frac{1}{2}\pi\varphi\right]\Big|_{t'}^{t''} = \\
&\quad = \int d^3k \Big[a^*(\boldsymbol{k})\, a(\boldsymbol{k}) - \frac{1}{2}(a^*(\boldsymbol{k}, t'')\, a(\boldsymbol{k}, t'') + \\
&\quad + a^*(\boldsymbol{k}, t')\, a(\boldsymbol{k}, t')) - (a(\boldsymbol{k}, t'') - a(\boldsymbol{k})\, e^{-i\omega t''})^2 - \\
&\qquad - (a^*(\boldsymbol{k}, t') - a^*(\boldsymbol{k})\, e^{i\omega t'})^2\Big]. \qquad (3.40)
\end{aligned}$$

At this point we shall stop transforming the integrand in (3.6) and pass over to the discussion of the limit process $t'' \to \infty$, $t' \to -\infty$. In the integral $\int_{t'}^{t''} dt \int d^3x\ \pi_1^2$ we can pass to the limit, if $\pi_1(\boldsymbol{x}, t)$ decreases at large t, so that

$$I(t) = \int \pi_1^2(\boldsymbol{x}, t)\, d^3x \tag{3.41}$$

is an integrable function of t as $|t| \to \infty$. In the following we shall call functions analogous to $\pi_1(\boldsymbol{x}, t)$ *rapidly decreasing.* The functions $\pi_1(\boldsymbol{x}, t)$ for which $\int_{-\infty}^{\infty} I(t)\, dt = \infty$ do not contribute to the S-matrix, if we stick to the convention that $\exp\{i\infty\} = 0$.

The boundary conditions (3.7), (3.8) define the asymptotic behavior of the variables $a^*(\boldsymbol{k}, t)$ as $t \to \infty$ and $a(\boldsymbol{k}, t)$ as $t \to -\infty$:

$$\begin{aligned} a^*(\boldsymbol{k}, t) &= a^*(\boldsymbol{k})\, e^{i\omega t} + a^*_{1,\,\text{out}}(\boldsymbol{k}, t), \quad t \to \infty, \\ a(\boldsymbol{k}, t) &= a(\boldsymbol{k})\, e^{-i\omega t} + a_{1,\,\text{in}}(\boldsymbol{k}, t), \quad t \to -\infty, \end{aligned} \tag{3.42}$$

where $a^*_{1,\,\text{out}}(\boldsymbol{k}, t)$ and $a_{1,\,\text{in}}(\boldsymbol{k}, t)$ are rapidly decreasing functions at $t \to \infty$ and $t \to -\infty$, respectively.

From (3.42) it follows that the differences

$$\partial_0 a^*(\boldsymbol{k}, t) - i\omega a^*(\boldsymbol{k}, t) \quad \text{and} \quad \partial_0 a(\boldsymbol{k}, t) + i\omega a(\boldsymbol{k}, t) \tag{3.43}$$

decrease rapidly as $t \to +\infty$ and $t \to -\infty$, respectively. Then, as is seen from (3.37), the difference $\pi - \partial_0\varphi = \pi_1$ will decrease rapidly as $|t| \to \infty$ only if

$$\begin{aligned} \partial_0 a(\boldsymbol{k}, t) + i\omega(\boldsymbol{k})\, a(\boldsymbol{k}, t) &= a_1(\boldsymbol{k}, t), \\ \partial_0 a^*(\boldsymbol{k}, t) - i\omega(\boldsymbol{k})\, a^*(\boldsymbol{k}, t) &= a_1^*(\boldsymbol{k}, t), \end{aligned} \tag{3.44}$$

where $a_1(\boldsymbol{k}, t)$ and $a_1^*(\boldsymbol{k}, t)$ derease rapidly as $t \to \infty$ and $t \to -\infty$.

The relation (3.44) together with (3.41) means that the integration variables $\varphi(\boldsymbol{x}, t)$ for $|t| \to \infty$ have asymptotic forms

$$\varphi(\boldsymbol{x}, t) = \varphi_{0,\,\substack{\text{in}\\ \text{out}}}(\boldsymbol{x}, t) + \varphi_{1,\,\substack{\text{in}\\ \text{out}}}(\boldsymbol{x}, t), \quad t \to \mp\infty, \tag{3.45}$$

where the $\varphi_{1,\,\mathrm{in}}(\boldsymbol{x}, t)$ decrease rapidly as $t \to -\infty$, the $\varphi_{1,\,\mathrm{out}}(\boldsymbol{x}, t)$ decrease rapidly as $t \to \infty$, and the $\varphi_{0,\,\substack{\mathrm{in}\\ \mathrm{out}}}(\boldsymbol{x}, t)$ are solutions of the free field equation of motion

$$\Box \varphi_{0,\,\substack{\mathrm{in}\\ \mathrm{out}}} + m^2 \varphi_{0,\,\substack{\mathrm{in}\\ \mathrm{out}}} = 0, \tag{3.46}$$

given by the formulas

$$\varphi_{0,\,\substack{\mathrm{in}\\ \mathrm{out}}}(x) = \\ = \left(\frac{1}{2\pi}\right)^{3/2} \int \left(a^*_{\substack{\mathrm{in}\\ \mathrm{out}}}(\boldsymbol{k})\, e^{ikx} + a_{\substack{\mathrm{in}\\ \mathrm{out}}}(\boldsymbol{k})\, e^{-ikx}\right)\Big|_{k_0=\omega} \frac{d^3k}{\sqrt{2\omega}}, \tag{3.47}$$

where

$$a_{\mathrm{in}}(\boldsymbol{k}) = a(\boldsymbol{k}), \qquad a^*_{\mathrm{out}}(\boldsymbol{k}) = a^*(\boldsymbol{k}). \tag{3.48}$$

No conditions are imposed on the functions $a_{\mathrm{out}}(\boldsymbol{k})$ and $a^*_{\mathrm{in}}(\boldsymbol{k})$.

If the integration variables behave asymptotically as described above, then the two last terms on the r.h.s. of (3.40) disappear in the limit $t'' \to -\infty$. Indeed, we have, for example,

$$\int d^3k\, (a^*(\boldsymbol{k}, t') - a^*(\boldsymbol{k})\, e^{i\omega t'})^2 = \int d^3k \big[(a^*_{\mathrm{in}}(\boldsymbol{k}) - a^*(\boldsymbol{k}))^2 e^{2i\omega t'} + \\ + 2\,(a^*_{\mathrm{in}}(\boldsymbol{k}) - a^*(\boldsymbol{k}))\, e^{i\omega t'} a_1(\boldsymbol{k}, t') + a_1^2(\boldsymbol{k}, t')\big]. \tag{3.49}$$

The last two terms here vanish as $t' \to -\infty$ owing to the rapid decrease of $a_1(\boldsymbol{k}, t)$, and the first one vanishes, according to the Riemann-Lebesgue lemma, owing to the oscillations of $\exp\{2i\omega t'\}$.

Let us now collect the contributions of the nonvanishing terms into the S-matrix. Note that the second term in (3.40) cancels out with the boundary terms of (3.6). As a result, we obtain for the kernel of the S-matrix the following expression:

$$S(a^*, a) = \exp\left\{ \int a^*(\boldsymbol{k})\, a(\boldsymbol{k})\, d^3k \right\} \int \exp\left\{ i \int dx \left[-\frac{1}{2}\pi_1^2 + \right.\right. \\ \left.\left. + \frac{1}{2}\partial_\mu \varphi_1 \partial_\mu \varphi_1 - \frac{m^2}{2}\varphi_1^2 - V(\varphi) \right] \right\} \prod_x \frac{d\varphi(x)\, d\pi(x)}{2\pi}, \tag{3.50}$$

where we have passed over to the relativistic notation $x = (\boldsymbol{x}, t)$. The variables $\varphi_1(x)$ and $\pi_1(x)$ are related to $\varphi(x)$ and $\pi(x)$ by the formulas (3.38).

In the formula (3.50) the variables π_1 and φ are completely separated, and we can integrate explicitly over π_1. The boundary conditions imposed on π_1 do not depend on $a^*(\boldsymbol{k})$ and $a(\boldsymbol{k})$; therefore, the integral

$$N^{-1} = \int \exp\left\{-i \int \frac{1}{2} \pi_1^2 \, dx\right\} \prod_x \frac{d\pi_1(x)}{2\pi} \tag{3.51}$$

represents just a normalizing constant.

The first factor in (3.50) may be dropped when passing from the kernel to the normal symbol of the S-matrix. As a result, the following explicitly relativistic expression is obtained for the normal symbol:

$$S(\varphi_0) = N^{-1} \int \exp\left\{ i \int dx \left[\frac{1}{2} \partial_\mu \varphi_1 \partial_\mu \varphi_1 - \right.\right.$$
$$\left.\left. - \frac{1}{2} m^2 \varphi_1^2 - V(\varphi) \right] \right\} \prod_x d\varphi(x), \tag{3.52}$$

where the integration runs over all fields $\varphi(x)$, which behave asymptotically in accordance with the formulas (3.45) to (3.48),

$$\varphi_1(x) = \varphi(x) - \varphi_0(x). \tag{3.53}$$

From comparison of (3.45) and (3.16) we see that φ_1 has the following asymptotic behavior:

$$\varphi_1(x) = \left(\frac{1}{2\pi}\right)^{3/2} \int b^*(\boldsymbol{k}) e^{ikx} \Big|_{k_0=\omega} \frac{d^3k}{\sqrt{2\omega}} + \varphi_{2,\,\mathrm{in}}(x), \quad t = -\infty,$$
$$\varphi_1(x) = \left(\frac{1}{2\pi}\right)^{3/2} \int b(\boldsymbol{k}) e^{-ikx} \Big|_{k_0=\omega} \frac{d^3k}{\sqrt{2\omega}} + \varphi_{2,\,\mathrm{out}}(x), \quad t \to \infty, \tag{3.54}$$

where the $\varphi_{2,\,\substack{\mathrm{in}\\ \mathrm{out}}}(x)$ decrease rapidly as $t \to \mp\infty$. We shall say that $\varphi_1(x)$ satisfies the Feynman radiation condition. In the alternative formulation $\varphi_1(x)$ has no incoming wave as $t \to -\infty$ and no outgoing wave as $t \to \infty$. The incoming and outgoing waves of the variable $\varphi(\boldsymbol{x}, t)$ are completely defined by the solution $\varphi_0(x)$.

When acting on the functions $\varphi_1(x)$ which satisfy the radiation condition, the Klein–Gordon operator $\Box + m^2$ is symmetric:

$$\int \varphi_1 (\Box + m^2) \varphi_1' \, dx = \int [(\Box + m^2) \varphi_1] \varphi_1' (x) \, dx = \\ = - \int (\partial_\mu \varphi_1 \partial_\mu \varphi_1' - m^2 \varphi_1 \varphi_1') \, dx. \quad (3.55)$$

Indeed, when the action of the operator $\Box$ is transferred to φ_1' from φ_1, then the terms outside of the integral, which appear when integration by parts is performed, take the form

$$\int d^3x \varphi_1 \partial_0 \varphi_1' \Big|_{t'}^{t''} = \\ = i \int d^3k \, [b^* (\boldsymbol{k}) \, b'^* (-\boldsymbol{k}) \, e^{2i\omega t''} - b (\boldsymbol{k}) \, b' (-\boldsymbol{k}) \, e^{-2i\omega t'}] + \ldots, \quad (3.56)$$

where .... stands for the terms containing $\varphi_{2,\, \substack{\text{in} \\ \text{out}}}$. The integrals (3.56) vanish as $t'' \to \infty$, $t' \to -\infty$ for the reasons which were described after (3.49). Thus, the quadratic form in (3.52) is defined uniquely as the quadratic form of the Klein–Gordon operator in the space of the functions $\varphi_1(x)$ which satisfy the radiation condition.

The action of the operator $\Box + m^2$ transforms the functions which satisfy the radiation condition into rapidly decreasing functions, and this action is reversible. The equation

$$(\Box + m^2) \varphi = \eta, \quad (3.57)$$

where φ satisfies the radiation condition, and η decreases rapidly, has a unique solution

$$\varphi (x) = \int D_c (x - y) \, \eta (y) \, dy, \quad (3.58)$$

the function $D_c(x)$ has been introduced previously in (3.18).

The formula (3.52) may be rewritten in a more explicit form:

$$S (\varphi_0) = N^{-1} \int\limits_{\varphi \to \varphi_{\substack{\text{in} \\ \text{out}}}} \exp \left\{ i \int \mathcal{L} (x) \, dx \right\} \prod_x d\varphi (x), \quad (3.59)$$

where it must be understood that the quadratic form in the action

$\int \mathscr{L}(x)\, dx$ is regularized so that

$$\int (\partial_\mu \varphi \partial_\mu \varphi - m^2 \varphi^2)\, dx = \\ = \int [\partial_\mu (\varphi - \varphi_0)\, \partial_\mu (\varphi - \varphi_0) - m^2 (\varphi - \varphi_0)^2]\, dx. \quad (3.60)$$

The left-hand side can be formally transformed into the right-hand one if one integrates by parts, forgetting about the terms outside of the integral.

The formula (3.59) may be taken as the starting point for the calculation of the S-matrix according to a scheme somewhat different from the above-described Feynman perturbation theory. This scheme is based on the formal application of the stationary-phase method to the integral (3.59) and is called the loop expansion. This method will not be discussed herein, and we shall confine ourselves to the Feynman perturbation theory.

Using (3.59), we can readily write down also the expression for the generating functional of the Green functions in terms of the path integral. Since $Z(\eta)$ is the transition amplitude from vacuum to vacuum in the presence of a source $\eta(x)$, we have

$$Z(\eta) = N^{-1} \int \exp\left\{ i \int [\mathscr{L}(x) + \eta(x)\, \varphi(x)]\, dx \right\} \prod_x d\varphi(x), \quad (3.61)$$

where integration is performed over the fields $\varphi(x)$ which satisfy the radiation condition.

The obtained formulas (3.59) and (3.61) are appealing in that they are compact and clear. For example, the representation of $Z(\eta)$ in the form of an integral allows one to use simple formulas of calculus: integration by parts, changing of the order of integration, substitution of variables, calculation by the stationary-phase method. Unfortunately, as has already been mentioned above, there does not exist at present a definition of this integral in internal terms, which would make all these formal transformations rigorous. Nevertheless, in the framework of perturbation theory a rigorous meaning may be attributed to the formula (3.61) by making use of (3.27), which expresses $Z(\eta)$ in terms of variational derivatives.

A rigorous proof based on this formula may be given in the framework of perturbation theory for all the abovementioned operations for the integral (3.61). This will be done in Section 2.5.

2.4. THE PATH INTEGRAL OVER FERMI FIELDS

The technique described in the previous sections can be applied practically without any changes to the case of several interacting scalar fields, and also to other Bose fields, including vector fields, which will be discussed in detail in the next chapter. In the present section we shall show that for Fermi fields it is possible to construct such an integration procedure, and that the corresponding dynamic formulas (the evolution operator, the S-matrix) will look practically the same as in the case of Bose fields.

Let us start with a Fermi system having one degree of freedom. The space of states of such a system is two-dimensional. In this space two operators $\boldsymbol{a}^*$ and $\boldsymbol{a}$ act which are complex conjugates of each other and satisfy commutation rules

$$\boldsymbol{a}^*\boldsymbol{a} + \boldsymbol{a}\boldsymbol{a}^* = 1, \qquad (\boldsymbol{a}^*)^2 = 0; \qquad \boldsymbol{a}^2 = 0. \tag{4.1}$$

These operators may be represented by 2×2 matrices

$$\boldsymbol{a}^* = \begin{pmatrix} 0 & 1 \\ 0 & 0 \end{pmatrix}; \quad \boldsymbol{a} = \begin{pmatrix} 0 & 0 \\ 1 & 0 \end{pmatrix}. \tag{4.2}$$

The formalism of path integration is based on another representation of the operators $\boldsymbol{a}^*$, $\boldsymbol{a}$, which is a particular analog of the holomorphic representation. Let us consider two anticommuting variables a^* and a

$$a^*a + aa^* = 0; \quad (a^*)^2 = 0; \quad a^2 = 0. \tag{4.3}$$

Such operators are called the generators of the Grassman algebra. The common element of this algebra (a function of the generators) is given by the formula

$$f(a^*, a) = f_{00} + f_{01}a + f_{10}a^* + f_{11}aa^*, \tag{4.4}$$

where f_{00}, f_{01}, f_{10}, f_{11} are complex numbers. We shall call functions depending only on a^* *holomorphic functions:*

$$f(a^*) = f_0 + f_1 a^*. \tag{4.5}$$

The set of such functions forms a two-dimensional space, and we shall use them for representation of the state vectors of our system.

We shall take the operators $\boldsymbol{a}^*$ and $\boldsymbol{a}$ in the form

$$\boldsymbol{a}^* f(a^*) = a^* f(a^*); \quad \boldsymbol{a} f(a^*) = \frac{d}{da^*} f(a^*), \tag{4.6}$$

where differentiation is defined naturally by the formula

$$\frac{d}{da^*}(f_0 + f_1 a^*) = f_1. \tag{4.7}$$

It is easy to verify that the commutation relations (4.1) are indeed satisfied. Our next task will be to introduce scalar product in the space of holomorphic functions such that $\boldsymbol{a}^*$ and $\boldsymbol{a}$ will be complex conjugates of each other. We shall do this by means of a convenient definition of the integral of functions of the form (4.4) over $da^*\, da$. We shall assume that da^* and da anticommute with each other as well as with a and a^*, and we define the following simple integrals:

$$\int a^*\, da^* = 1; \quad \int a\, da = 1; \quad \int da^* = 0; \quad \int da = 0. \tag{4.8}$$

We may point out that the explicit meaning of the last two formulas consists in the fact that the integral of a total derivative is equal to zero. The said rules are sufficient for defining an integral of any function, if we also stick to the convention that a multiple integral is understood to be a repeated integral. Then

$$\int f(a^*, a)\, da^*\, da = f_{11}. \tag{4.9}$$

The scalar product to be found is given by the formula

$$(f_1, f_2) = \int (f_1(a^*))^* f_2(a^*)\, e^{-a^* a}\, da^*\, da; \tag{4.10}$$

here it is to be understood that

$$(f(a^*))^* = f_0^* + f_1^* a. \tag{4.11}$$

Let us check that this scalar product is positive definite. For this purpose, we shall show that the monomials

$$\psi_0 = 1; \quad \psi_1 = a^* \tag{4.12}$$

are orthonormalized. We have

$$(\psi_0, \psi_0) = \int e^{-a^*a}\, da^*\, da = \int (1 - a^*a)\, da^*\, da = 1; \quad (4.13)$$

$$(\psi_0, \psi_1) = \int a^* e^{-a^*a}\, da^*\, da = 0; \quad (4.14)$$

$$(\psi_1, \psi_1) = \int a a^* e^{-a^*a}\, da^*\, da = 1. \quad (4.15)$$

That the operators $\boldsymbol{a}^*$, $\boldsymbol{a}$ are conjugates of each other follows from the fact that they are given by the matrices (4.2) in the basis ψ_0, ψ_1. Indeed,

$$\begin{aligned} \boldsymbol{a}^*\psi_0 &= \psi_1; \quad \boldsymbol{a}^*\psi_1 = 0; \\ \boldsymbol{a}\psi_0 &= 0; \quad \boldsymbol{a}\psi_1 = \psi_0. \end{aligned} \quad (4.16)$$

Let us apply the integration rules formulated above to calculate the integral of an exponential function whose argument is a nonhomogeneous quadratic form:

$$\int \exp\{a^*Aa + a^*b + b^*a\}\, da^*\, da, \quad (4.17)$$

where b and b^* anticommute between themselves and a^*, a.

As follows from (4.9), we can shift the integration variables in the integral (4.17):

$$a^* \to a^* - A^{-1}b^*; \quad a \to a - A^{-1}b, \quad (4.18)$$

since the coefficient of aa^* in the integrand does not change under such a shift. After this shift the integral (4.17) acquires the form

$$\exp\{-b^*A^{-1}b\} \int \exp\{a^*Aa\}\, da^*\, da = \\ = -A \exp\{-b^*A^{-1}b\}. \quad (4.19)$$

Note that the formula (4.19) looks exactly as in the case of integration over commuting variables, except that the factor is in the numerator, instead of in the denominator as it would be in the case of commuting variables.

We shall now proceed to describe methods of defining operators in the representation under consideration. An operator of the general form may be determined as

$$A = K_{00} + K_{10}\boldsymbol{a}^* + K_{01}\boldsymbol{a} + K_{11}\boldsymbol{a}^*\boldsymbol{a}. \tag{4.20}$$

Two functions can be associated with it on the Grassman algebra: the normal symbol

$$K(a^*, a) = K_{00} + K_{10}a^* + K_{01}a + K_{11}a^*a, \tag{4.21}$$

and the kernel

$$A(a^*, a) = A_{00} + A_{10}a^* + A_{01}a + A_{11}a^*a, \tag{4.22}$$

where $A_{n,m}$, $n, m = 0, 1$, are the matrix elements of the operator A in the base ψ_0, ψ_1:

$$A_{nm} = \langle \psi_n \mid A \mid \psi_m \rangle. \tag{4.23}$$

It is obvious that

$$(Af)(a^*) = \int A(a^*, \alpha) f(\alpha^*) e^{-\alpha^*\alpha} d\alpha^* d\alpha; \tag{4.24}$$

$$(A_1A_2)(a^*, a) = \int A_1(a^*, \alpha) A_2(\alpha^*, a) e^{-\alpha^*\alpha} d\alpha^* d\alpha. \tag{4.25}$$

In order to write down these formulas, we have had to introduce new anticommuting variables α^*, α. By definition, α^*, α anticommute with a^*, a.

The normal symbol $K(a^*, a)$ and the kernel $A(a^*, a)$ of the given operator A are related by the formula

$$A(a^*, a) = e^{a^*a} K(a^*, a). \tag{4.26}$$

To prove this statement, it is sufficient to compare the coefficients K_{nm} and A_{nm} in the formulas (4.21) and (4.22) and verify that

$$K_{00} = A_{00}; \quad K_{01} = A_{01}; \quad K_{10} = A_{10}; \quad K_{11} = A_{11} - A_{00}. \tag{4.27}$$

All the above formulas are readily generalized to the case of n degrees of freedom. For this purpose one must use $2n$ anticommuting variables

$$a_1, \ldots, a_n; \quad a_1^*, \ldots, a_n^*. \tag{4.28}$$

The space of state vectors consists of analytic functions $f(a^*)$ and has dimension equal to 2^n. The operators $\boldsymbol{a}_i^*, \boldsymbol{a}_i$, $i = 1, \ldots, n$, act according to the rule

$$\boldsymbol{a}_i f(a^*) = \left(\frac{\partial}{\partial a_i^*}\right)_L f(a^*); \quad \boldsymbol{a}_i^* f(a^*) = a_i^* f(a^*), \tag{4.29}$$

where the subscript L signifies that in differentiating with respect to a_i^* in the function $f(a^*)$ we must displace the variable a_i^* to the left before dropping it.

The operators introduced satisfy the commutation relations

$$\boldsymbol{a}_i^* \boldsymbol{a}_k + \boldsymbol{a}_k \boldsymbol{a}_i^* = \delta_{ik}; \quad \boldsymbol{a}_i^* \boldsymbol{a}_k^* + \boldsymbol{a}_k^* \boldsymbol{a}_i^* = 0; \quad \boldsymbol{a}_i \boldsymbol{a}_k + \boldsymbol{a}_k \boldsymbol{a}_i = 0 \tag{4.30}$$

and are conjugates of each other with respect to the scalar product

$$(f_1, f_2) = \int (f_1(a^*))^* f_2(a^*) e^{-\Sigma a^* a} \prod da^* \, da. \tag{4.31}$$

Here the operation * is defined by the formula

$$(C a_{i_1}^* \ldots a_{i_r}^*)^* = C^* a_{i_r} \ldots a_{i_1}; \tag{4.32}$$

and the integration is performed as before.

The integral of an arbitrary function $f(a^*, a)$ equals

$$\int f(a^*, a) \prod da^* \, da = f_{1, \ldots, n, n, \ldots, 1}, \tag{4.33}$$

where $f_{1,\ldots,n,n,\ldots,1}$ is the coefficient of the monomial $a_1 \cdots a_n a_n^* \cdots a_1$ in the expansion of f in the generators. The Gaussian integral

$$\int \exp\{a_i^* A_{ik} a_k + a_i^* b_i + b_i^* a_i\} \prod_i da_i^* \, da_i \tag{4.34}$$

is calculated by shifting, as in the case of one degree of freedom, and is equal to

$$\exp\{- b_i^* (A^{-1})_{ik} b_k\} \int \exp\{a_i^* A_{ik} a_k\} \prod_i da_i^* \, da_i. \tag{4.35}$$

The remaining integral, in view of (4.33), is equal to det $A(-1)^n$. We may point out that the exponential function in the answer may be calculated by substituting into the integrand the solution of the equations

$$\left(\frac{d}{da_i^*}\right)_L (a_i^* A_{ik} a_k + a_i^* b_i + b_i^* a_i) = 0;$$
$$\left(\frac{d}{da_i}\right)_L (a_i^* A_{ik} a_k + a_i^* b_i + b_i^* a_i) = 0. \tag{4.36}$$

This property is general for Gaussian integrals both over usual commuting and over anticommuting variables. We shall frequently use it in the future. The monomials

$$\psi_{i_1, \ldots, i_r} = a_{i_1}^* \ldots a_{i_r}^* \qquad (i_1 < i_2 < \ldots < i_r) \tag{4.37}$$

are orthonormalized and constitute a basis in the space of states.

As in the case of one degree of freedom, an arbitrary operator A can be expressed by the normal symbol $K(a^*, a)$ or by the kernel $A(a^*, a)$. If the operator A is given by the expression

$$A = \sum_{r,t} \sum_{\substack{i_1 < \ldots < i_r \\ j_1 < \ldots < j_t}} K_{i_1 \ldots i_r | j_1 \ldots j_t} \boldsymbol{a}_{i_1}^* \ldots \boldsymbol{a}_{i_r}^* \boldsymbol{a}_{j_1} \ldots \boldsymbol{a}_{j_t}, \tag{4.38}$$

then

$$K(a^*, a) = \sum_{r,t} \sum_{\substack{i_1 < \ldots < i_r \\ j_1 < \ldots < j_t}} K_{i_1 \ldots i_r | j_1 \ldots j_t} a_{i_1}^* \ldots a_{i_r}^* a_{j_1} \ldots a_{j_t} \tag{4.39}$$

and

$$A(a^*, a) =$$
$$= \sum_{r,t} \sum_{\substack{i_1 < \ldots < i_r \\ j_1 < \ldots < j_t}} A_{i_1 \ldots i_r | j_1 \ldots j_t} a_{i_1}^* \ldots a_{i_r}^* a_{j_1} \ldots a_{j_t}, \tag{4.40}$$

where

$$A_{i_1 \ldots i_r | j_1 \ldots j_t} = \langle \psi_{i_1 \ldots i_r} | A | \psi_{j_1 \ldots j_t} \rangle. \tag{4.41}$$

The kernel and the normal symbol are related to each other by

$$A(a^*, a) = e^{\sum a_i^* a_i} K(a^*, a). \tag{4.42}$$

The action of the operator on a function, and the product of operators, are given by the formulas

$$(Af)(a^*) = \int A(a^*, \alpha) f(\alpha^*) e^{-\sum \alpha^* \alpha} \prod d\alpha^* \, d\alpha; \tag{4.43}$$

$$(A_1 A_2)(a^*, a) = \int A_1(a^*, \alpha) A_2(\alpha^*, a) e^{-\sum \alpha^* \alpha} \prod d\alpha^* \, d\alpha. \tag{4.44}$$

Comparison of the above formulas with the formulas (2.11), (2.14), derived in Section 2.2 for Bose systems, reveals that they have the same form. By following the derivation of the representation for the kernel of the evolution operator in terms of the path integral, one may verify that this derivation is based entirely on the two formulas (2.17) and (2.14). In the case of fermions we have absolutely identical formulas (4.42) and (4.44). Therefore, the representation for the kernel of the evolution operator for a Fermi system with the Hamiltonian $h(\boldsymbol{a}^*, \boldsymbol{a}, t)$ can be immediately written down as

$$U(a^*, a; t'', t') = \\ = \int \exp\left\{ \frac{1}{2} \sum_k \left(a_k^*(t'') a_k(t'') + a_k^*(t') a_k(t') \right) + \right. \\ \left. + i \int_{t'}^{t''} \left[\frac{1}{2i} \sum_k \left(a_k^* \dot{a}_k - \dot{a}_k^* a_k \right) - h(a^*(t), a(t), t) \right] dt \right\} \prod_{t,\,k} da^* \, da, \tag{4.45}$$

where we assume that

$$a_k^*(t'') = a_k^*; \quad a_k(t') = a_k. \tag{4.46}$$

We emphasize that we are dealing here with an integral over the infinite-dimensional Grassman algebra with independent generators $a_k^*(t)$, $a_k(t)$, $k = 1, \ldots, n$ for each t, $t' \le t \le t''$.

Let us pass now to the field theory. A complex spinor field may be regarded as a system of fermions with an infinite number of degrees of freedom. In this case the generators of the Grassman algebra are the anticommuting functions $\psi(\boldsymbol{x})$, $\bar{\psi}(\boldsymbol{x})$ or the functions $b_i(\boldsymbol{k})$, $b_i^*(\boldsymbol{k})$, $c_i(\boldsymbol{k})$, $c_i^*(\boldsymbol{k})$, $i = 1, 2$, which are linearly related to them:

$$\psi(\boldsymbol{x}) = \left(\frac{1}{2\pi}\right)^{3/2} \int \left(e^{-i\boldsymbol{k}\boldsymbol{x}} u_i(\boldsymbol{k})\, b_i^*(\boldsymbol{k}) + e^{i\boldsymbol{k}\boldsymbol{x}} v_i(\boldsymbol{k})\, c_i(\boldsymbol{k})\right) d^3k,$$

$$\psi^*(\boldsymbol{x}) = \left(\frac{1}{2\pi}\right)^{3/2} \int \left(e^{i\boldsymbol{k}\boldsymbol{x}} u_i^*(\boldsymbol{k})\, b_i(\boldsymbol{k}) + e^{-i\boldsymbol{k}\boldsymbol{x}} v_i^*(\boldsymbol{k})\, c_i^*(\boldsymbol{k})\right) d^3k, \tag{4.47}$$

where $u_i(\boldsymbol{k})$, $v_i(\boldsymbol{k})$ are two pairs of linearly independent solutions of the Dirac equation

$$\begin{aligned} (\gamma_\mu k_\mu - m)\, u_i(\boldsymbol{k}) \big|_{k_0 = \sqrt{\boldsymbol{k}^2 + m^2}} &= 0; \\ (\gamma_\mu k_\mu - m)\, v_i(\boldsymbol{k}) \big|_{k_0 = -\sqrt{\boldsymbol{k}^2 + m^2}} &= 0; \end{aligned} \qquad i = 1, 2. \tag{4.48}$$

In terms of the generators b, b^*, c, c^* the normal product is defined in the usual way, that is, in the expressions for arbitrary operators in terms of the kernel $A(b^*, c^*; b, c)$ or of the normal symbol $K(b^*, c^*; b, c)$ the generators b^*, c^* are placed to the left of b, c.

Let us consider a system of Fermi fields interacting with an external source. As an external source we shall take anticommuting spinor functions $\xi(x)$, $\bar{\xi}(x)$. The Hamiltonian of such a system has the form

$$\begin{aligned} h = \int \big(i\bar{\psi}(\boldsymbol{x})\, \gamma_k \partial_k \psi(\boldsymbol{x}) + m\bar{\psi}(\boldsymbol{x})\, \psi(\boldsymbol{x}) + \bar{\psi}(\boldsymbol{x})\, \xi(x) + {} \\ + \bar{\xi}(x)\, \psi(\boldsymbol{x})\big)\, d^3x = \int \big[\sqrt{\boldsymbol{k}^2 + m^2}\, \big(b_i^*(\boldsymbol{k})\, b_i(\boldsymbol{k}) + c_i^*(\boldsymbol{k})\, c_i(\boldsymbol{k})\big) + {} \\ + \gamma_i^*(\boldsymbol{k}, t)\, b_i + b_i^* \gamma_i(\boldsymbol{k}, t) + \delta_i^*(\boldsymbol{k}, t)\, c_i(\boldsymbol{k}) + c_i^*(\boldsymbol{k})\, \delta_i(\boldsymbol{k}, t)\big]\, d^3k. \end{aligned} \tag{4.49}$$

Here we have used the orthonormality properties of the spinors u_i and v_i, $i = 1, 2$, in passing to the momentum representation. Then

$$\begin{aligned} \gamma_i(\boldsymbol{k}, t) = u_i^* \tilde{\xi}(\boldsymbol{k}, t); \quad \delta_i(\boldsymbol{k}, t) = v_i^* \tilde{\xi}(\boldsymbol{k}, t), \\ \tilde{\xi}(\boldsymbol{k}, t) = \left(\frac{1}{2\pi}\right)^{3/2} \int \xi(\boldsymbol{x}, t)\, e^{i\boldsymbol{k}\boldsymbol{x}}\, d^3x. \end{aligned} \tag{4.50}$$

The S-matrix in the form of a path integral is given by the formula

$$S(b^*, c^*; b, c) = \lim_{\substack{t'' \to \infty \\ t' \to -\infty}} \int \exp\Big\{ \frac{1}{2} \int d^3k \,(b_i^*(\boldsymbol{k}, t'')\, b_i(\boldsymbol{k}, t'') +$$

$$+ b_i^*(\boldsymbol{k}, t')\, b_i(\boldsymbol{k}, t') + (b \leftrightarrow c) +$$

$$+ i \int_{t'}^{t''} dt \Big[\Big[\int d^3k \frac{1}{2i} (\dot{b}_i^*(\boldsymbol{k}, t)\, b_i(\boldsymbol{k}, t) - b_i^*(\boldsymbol{k}, t)\, \dot{b}_i(\boldsymbol{k}, t) +$$

$$+ (b \leftrightarrow c)\big) - h(b^*, b, c^*, c)\Big]\Big\}, \qquad (4.51)$$

which is Gaussian and is calculated exactly in the same way as the corresponding integral (3.6), (3.15) for the scalar field. The expression for the normal symbol is written down in the explicitly relativistic form

$$S_{\text{norm}}(\bar{\xi}, \xi; b^*, b, c^*, c) = \exp\Big\{ i \int \bar{\xi}(x)\, S_c(x-y)\, \xi(y)\, dx\, dy +$$

$$+ i \int (\bar{\xi}(x)\, \psi_0(x) + \bar{\psi}_0(x)\, \xi(x))\, dx \Big\}, \qquad (4.52)$$

where

$$S_c(x-y) = \left(\frac{1}{2\pi}\right)^3 \frac{1}{i} \int e^{ik(x-y)} e^{-ik_0|x_0-y_0|} \big(v_i(\boldsymbol{k}) \otimes v_i^*(\boldsymbol{k}) +$$

$$+ u_i(\boldsymbol{k}) \otimes u_i^*(\boldsymbol{k})\big)\, d^3k =$$

$$= -\left(\frac{1}{2\pi}\right)^4 \int (\gamma_\mu k_\mu - m + i0)^{-1} e^{-ik(x-y)} d^4k \qquad (4.53)$$

is the causal Green function of the Dirac equation and

$$\psi_0(x) = \left(\frac{1}{2\pi}\right)^{3/2} \int \big(b_i^*(\boldsymbol{k})\, u_i(\boldsymbol{k})\, e^{ikx} + v_i(\boldsymbol{k})\, c_i(\boldsymbol{k})\, e^{-ikx}\big)\big|_{k_0=\omega}\, d^3k \qquad (4.54)$$

is the solution of the free Dirac equation

$$(i\gamma_\mu \partial_\mu - m)\, S_c(x) = \delta(x), \quad (i\gamma_\mu \partial_\mu - m)\, \psi_0 = 0. \qquad (4.55)$$

In order to pass from the first representation for S_c to the second one, the properties of the spinors u_i, v_i are again used. The first representation makes it obvious in what sense the Green function S_c is causal:

$$\psi_1(x) = \int S_c(x-y)\,\xi(y)\,d^4y \tag{4.56}$$

has no incoming (outgoing) waves as $t \to -\infty$ ($t \to \infty$).

The formula (4.52) can be taken as the basis for the derivation of the perturbation-theory expansion for the S-matrix of the spinor field, interacting with itself or with other fields. For this we point out that

$$\begin{aligned}
&\left(\frac{1}{i}\frac{\delta}{\delta\bar{\xi}(x)}\right)_L \exp\left\{i\int(\bar{\xi}\psi+\bar{\psi}\xi)\,dx\right\} = \\
&\qquad\qquad = \psi(x)\exp\left\{i\int(\bar{\xi}\psi+\bar{\psi}\xi)\,dx\right\}, \\
&\left(\frac{1}{i}\frac{\delta}{\delta\xi(x)}\right)_R \exp\left\{i\int(\bar{\xi}\psi+\bar{\psi}\xi)\,dx\right\} = \\
&\qquad\qquad = \exp\left\{i\int(\bar{\xi}\psi+\bar{\psi}\xi)\,dx\right\}\bar{\psi}(x),
\end{aligned} \tag{4.57}$$

where the definition of the right-hand derivative modifies in a natural way the definition of the left-hand one. These formulas together with (4.52) allow one to reduce the path integral for the S-matrix with an arbitrary interaction of the fields $\bar{\psi}$, ψ, φ to the integral for the S-matrix with an external source. The Green functions of the spinor field and the reduction formulas are obtained by a natural modification of the formula for the scalar field. We shall conclude our discussion by giving the expression for the generating functional for the Green functions for interacting spinor and scalar fields with the Lagrangian

$$\begin{aligned}
\mathcal{L}(x) = \bar{\psi}(x)\,i\gamma_\mu\partial_\mu\psi(x) - m\bar{\psi}(x)\,\psi(x) + \frac{1}{2}\partial_\mu\varphi(x)\,\partial_\mu\varphi(x) - \\
- \frac{1}{2}m^2\varphi^2(x) - g\bar{\psi}(x)\,\psi(x)\,\varphi(x),
\end{aligned} \tag{4.58}$$

containing the most simple version of interaction; this functional is given by the expression

$$\begin{aligned}
Z(\eta, \bar{\xi}, \xi) = \exp\left\{-ig\int\frac{1}{i}\left(\frac{\delta}{\delta\bar{\xi}(x)}\frac{\delta}{\delta\xi(x)}\frac{\delta}{\delta\eta(x)}\right)dx\right\}\times \\
\times\exp\left\{i\int\left(\bar{\xi}(x)\,S_c(x-y)\,\xi(y) + \frac{1}{2}\eta(x)\,D(x-y)\,\eta(y)\right)dx\,dy\right\}
\end{aligned} \tag{4.59}$$

and can be written down in the form of the path integral

$$Z(\eta, \bar{\xi}, \xi) = \\ = \frac{1}{N} \int \exp\left\{ i \int (\mathscr{L}(x) + \eta\varphi + \bar{\xi}\psi + \bar{\psi}\xi)\, dx \right\} \prod_x d\bar{\psi}\, d\psi\, d\varphi, \tag{4.60}$$

where the integration runs over the fields $\bar{\psi}(x)$, $\psi(x)$, $\varphi(x)$ which do not have incoming (outgoing) waves asymptotically as $t \to -\infty$ $(t \to \infty)$.

2.5. THE PROPERTIES OF THE PATH INTEGRAL IN PERTURBATION THEORY

As we already mentioned above, at present there is no definition of the path integral in internal terms. However, for the purposes of perturbation theory in quantum field theory it is sufficient to be able to work with path integrals of a special type—namely, with Gaussian integrals. For such integrals it is possible to develop a calculation and transformation technique, which contains in a compact and clear form all the combinatorics of the diagram technique of perturbation theory.

Let us obtain these rules in the case of a scalar field, taking for an example the generating functional for the Green functions. For this functional we have two equivalent representations: in the form of a path integral (3.61) and the explicit formula (3.27). We shall use the formula (3.27) as the definition of the Gaussian path integral. More precisely, we shall assume that

$$\int \exp\left\{ i \int \frac{1}{2} \varphi(x) K(x-y) \varphi(y)\, dx\, dy + i \int \varphi(x) \eta(x)\, dx \right\} \times \\ \times \varphi(x_1) \ldots \varphi(x_n) \prod_x d\varphi(x) = (-i)^n \frac{\delta}{\delta\eta(x_1)} \ldots \\ \ldots \frac{\delta}{\delta\eta(x_n)} \exp\left\{ -\frac{i}{2} \int \eta(x) K^{-1}(x-y) \eta(y)\, dx\, dy \right\}. \tag{5.1}$$

By definition, integration over $d\varphi$ is interchangeable with integration over dx and with differentiation with respect to external sources $\eta(x)$. It is assumed that an operator K with a kernel $K(x-y)$ has an inverse K^{-1} with a kernel $K^{-1}(x-y)$:

$$\int K(x-z) K^{-1}(z-y)\, dz = \\ = \int K^{-1}(x-z) K(z-y)\, dz = \delta(x-y). \tag{5.2}$$

We shall consider the kernel $K^{-1}(x-y)$ to be a sufficiently smooth function. The function $D_c(x-y)$ which enters into the generating functional $Z(\eta)$ does not, of course, have this property; this leads to the appearance of ultraviolet divergences when the functional is calculated. These divergences are removed by the renormalization procedure, which will be discussed in Chapter Four. The first step in this procedure consists of introducing an intermediate regularization, which substitutes the function $D_c(x-y)$ by a smooth function. Hence, the reasoning which now follows has to do with the regularized perturbation theory.

The class of functions $\varphi(x)$ over which integration is performed must provide a unique definition of the operator inverse to K. If $K = \Box$, then such a condition is the already mentioned causality condition: φ behaves asymptotically for $|t| \to \infty$ as the solution of the free equation which has no incoming waves as $t \to -\infty$ and no outgoing waves as $t \to \infty$. The operator K^{-1} in this case is integral with the kernel D_c (more precisely, as just mentioned, with its regularization). We shall call the integrand in this formula (5.1) at $\eta = 0$ a Gaussian functional.

We shall now pass to the discussion of the path integral defined by the formula (5.1).

First, let us point out that the functional (5.1), which it is natural to call the Fourier transform of the Gaussian functional

$$F(\varphi) = \exp\left\{ i \int \frac{1}{2} \varphi(x) K(x-y) \varphi(y) \, dx \, dy \right\} \varphi(x_1) \ldots \varphi(x_n), \tag{5.3}$$

is itself a Gaussian functional, since by differentiating $\exp\{\int \eta K^{-1} \eta \, dx \, dy\}$, we obtain an expression in which this exponential function is multiplied by a polynomial.

We shall show that our definition allows one to prove for the integral (5.1) the validity of the simplest transformations such as integration by parts and change of variables, and also to introduce the concept of the functional δ-function.

1. Integration by parts. Let us consider the integral

$$I = \int \left[\frac{\delta}{\delta\varphi(z)} \exp\left\{ \frac{i}{2} \int \varphi(x) K(x-y) \varphi(y) \, dx \, dy \right\} \right] \times \\ \times \exp\left\{ i \int \varphi(x) \eta(x) \, dx \right\} \prod_x d\varphi. \tag{5.4}$$

The functional

$$\frac{\delta}{\delta\varphi(z)} \exp\left\{\frac{i}{2}\int \varphi(x) K(x-y) \varphi(y)\, dx\, dy\right\} \tag{5.5}$$

is Gaussian; therefore the integral (5.4) makes sense and by definition is equal to

$$I = i\int\Big[\int K(z-y)\varphi(y)\,dy \times$$
$$\times \exp\left\{\frac{i}{2}\int \varphi(x) K(x-y)\varphi(y)\,dx\,dy + i\int\varphi(x)\eta(x)\,dx\right\}\Big]\prod_x d\varphi =$$
$$= \int K(z-y)\frac{\delta}{\delta\eta(y)} Z(\eta)\,dy = -i\eta(z) Z(\eta). \tag{5.6}$$

On the other hand

$$-i\eta(z) Z(\eta) = -\int \exp\left\{\frac{i}{2}\int \varphi(x) K(x-y)\varphi(y)\,dx\,dy\right\}\times$$
$$\times \frac{\delta}{\delta\varphi(z)} \exp\left\{i\int \varphi(x)\eta(x)\,dx\right\}\prod_x d\varphi. \tag{5.7}$$

Comparing (5.4) and (5.6), we see that we have the usual formula for integration by parts, and the boundary terms are dropped. This result is generalized in an obvious way to an arbitrary Gaussian integral, since any such integral may be represented as the derivative of I with respect to η.

2. Repeated integrals. Since an integral of a Gaussian functional is itself a Gaussian functional, it is possible to define a repeated integral. Let us show that

$$\int \exp\left\{ i \sum_{i,j=1}^{n} \frac{(K_n)_{ij}^{xy}}{2} \varphi_i^x \varphi_j^y + \sum_{j=1}^{n} \eta_j^x \varphi_j^x \right\} \prod_x d\varphi_1 \ldots d\varphi_n =$$
$$= \exp\left\{-\frac{i}{2}\sum_{i,j=1}^{n} (K_n^{-1})_{ij}^{xy}\eta_i^x\eta_j^y\right\} \tag{5.8}$$

(we use here abbreviated notation, having in mind that integration is performed over continuous indices x, y).

Let the equality (5.8) be valid for some n. We shall prove that it is valid also for the number $n+1$. According to our assumption

$$I_{n+1}=\int \exp\left\{-i\sum_{i,\,j=1}^{n}\left(\eta_i^x+K_{i\,n+1}^{xy}\varphi_{n+1}^y\right)\times\right.$$
$$\times\frac{\left(K_n^{-1}\right)_{ij}^{xz}}{2}\left(\eta_j^z+K_{j\,n+1}^{zu}\varphi_{n+1}^u\right)+$$
$$\left.+\frac{i}{2}K_{n+1\,n+1}^{xy}\varphi_{n+1}^x\varphi_{n+1}^y+i\eta_{n+1}^x\varphi_{n+1}^x\right\}\prod_x d\varphi_{n+1}. \tag{5.9}$$

Integrating over φ_{n+1}, we obtain

$$I_{n+1}=\exp\left\{-\frac{i}{2}\left(\eta_{n+1}^x-\sum_{i,\,j=1}^{n}K_{i\,n+1}^{xy}\left(K_n^{-1}\right)_{ij}^{yz}\eta_j^z\right)\times\right.$$
$$\times\left(K_{n+1\,n+1}^{zu}-\sum_{l,\,m=1}^{n}K_{l\,n+1}^{zv}\left(K_n^{-1}\right)_{lm}^{vs}K_{m\,n+1}^{su}\right)^{-1}\times$$
$$\left.\times\left(\eta_{n+1}^u-\sum_{i,\,j=1}^{n}K_{i\,n+1}^{ur}\left(K_n^{-1}\right)_{ij}^{rt}\eta_j^t\right)-\frac{i}{2}\sum_{i,\,j=1}^{n}\eta_i^x\left(K_n^{-1}\right)_{ij}^{xy}\eta_j^y\right\}. \tag{5.10}$$

Taking advantage of

$$\left(K^{-1}\right)_{ij}^{xy}=\left(\det K^{-1}\right)^{xz}\bar{K}_{ij}^{zy}, \tag{5.11}$$

where $\bar{K}_{ij}$ stands for the adjoint of the K_{ij}th element of the matrix K, one can represent the second factor in the power of the exponential function as

$$\left(\det K_n\right)^{zx}\left[\left(\det K_n\right)^{xy}K_{n+1\,n+1}^{yu}-\sum_{i,\,j=1}^{n}K_{i\,n+1}^{xy}\bar{K}_{ij}^{yv}K_{j\,n+1}^{vu}\right]^{-1}=$$
$$=\left(\det K_n\right)^{zx}\left(\det K_{n+1}^{-1}\right)^{xu}. \tag{5.12}$$

Let us consider the separate terms in the formula (5.10):

$$\eta_{n+1}^x\eta_{n+1}^y\left(\det K_n\right)^{xz}\left(\det K_{n+1}^{-1}\right)^{zy}=\eta_{n+1}^x\left(K_{n+1}^{-1}\right)_{n+1\,n+1}^{xy}\eta_{n+1}^y,$$
$$\sum_i\eta_{n+1}^x\left(\det K_{n+1}^{-1}\right)^{xy}K_{i\,n+1}^{yz}\bar{K}_{ij}^{zu}\eta_j^u=$$

$$=\eta^{x}_{n+1}\left(\det K^{-1}_{n+1}\right)^{xy}\bar{K}^{yz}_{n+1\,j}\eta^{z}_{j}=\eta^{x}_{n+1}\left(K^{-1}_{n+1}\right)^{xy}_{n+1\,j}\eta^{y}_{j}. \tag{5.13}$$

In an analogous manner it is possible to show that the coefficients of $\eta_i\eta_j$ are $(K^{-1}{}_{n+1})^{xy}_{ij}$. As a result, we obtain

$$I_{n+1}=\exp\left\{-\frac{i}{2}\sum_{i,\,j=1}\left(K^{-1}_{n+1}\right)^{xy}_{ij}\eta^{x}_{i}\eta^{y}_{j}\right\}, \tag{5.14}$$

which was to be proved. Obviously, the result does not depend on the order of integration, since a change in the order is just equivalent to the rearrangement of the columns of the matrix K. Therefore, it is proved that the repeated integrals exist, and the result does not depend on the order of integration.

3. Definition of the δ-function.

$$\int\exp\left\{i\int\eta(x)\left[\int c(x-y)\,\varphi(y)\,dy-\varphi'(x)\right]dx\right\}\prod_x d\eta \overset{\text{def}}{=\!=}$$

$$\overset{\text{def}}{=\!=}\delta\left(\varphi(x)-\int c^{-1}(x-y)\,\varphi'(y)\,dy\right). \tag{5.15}$$

This equality means that

$$\int F(\varphi)\left[\int\exp\left\{i\int\eta(x)\left[\int c(x-y)\,\varphi(y)\,dy-\varphi'(x)\right]dx\right\}\times\right.$$

$$\left.\times\prod_x d\eta\right]\prod_x d\varphi \overset{\text{def}}{=\!=} \int\left[\int F(\varphi)\exp\left\{i\int\eta(x)\left[\int c(x-y)\,\varphi(y)-\right.\right.\right.$$

$$\left.\left.-\varphi'(x)\right]dx\right\}\prod_x d\varphi\prod_x d\eta=F(c^{-1}\varphi'), \tag{5.16}$$

where $F(\varphi)$ is a Gaussian functional. [By analogy to the usual definition of the δ-function, one might expect that in (5.16) there should be still another constant (that is, independent of φ') factor $\det c^{-1}$. The absence of this factor is explained by the fact that our definition of a functional integral (5.1) includes the normalization condition.]

The equality (5.16) is verified by direct calculation:

$$\int\Big[\int \exp\left\{\frac{i}{2}\int \varphi(x) K(x-y)\varphi(y)\,dx\,dy\right\}\times$$
$$\times\exp\left\{i\int\eta(x)\Big[\int c(x-y)\varphi(y)dy-\varphi'(x)\Big]dx\right\}\prod_x d\varphi\prod_x d\eta=$$
$$=\exp\left\{\frac{i}{2}\int\varphi'(x)\,c^{-1}K(x-y)\,c^{-1}\varphi'(y)\,dx\,dy\right\}. \qquad (5.17)$$

We shall show now that

$$\int\exp\left\{i\int[f_x(\varphi)-\varphi'(x)]\,\eta(x)\,dx\right\}\prod_x d\eta=\delta(f_x(\varphi)-\varphi'(x)), \qquad (5.18)$$

$f_x(\varphi)$ as a function of x belongs to the same class as $\varphi(x)$. The function $f_x(\varphi)$ can be expanded in a formal series of the form

$$f_x(\varphi)=c_0(x)+\varphi(x)+\tilde f(\varphi),$$
$$\tilde f(\varphi)=g\int c_1(x,\ y)\,\varphi(y)\,dy+$$
$$+g^2\int c_2(x,\ y,\ z)\,\varphi(y)\,\varphi(z)\,dy\,dz+\ \dots \qquad (5.19)$$

(For simplicity we assume the coefficient of the first power of φ to be equal to 1. This reasoning is trivially generalized to the case of $c\neq 1$ with the help of the previous formula.) The equation

$$c_0(x)+\varphi(x)+\tilde f(\varphi)-\varphi'(x)=0 \qquad (5.20)$$

has a unique solution, which may be represented as a formal series in g.

Equation (5.18) means that

$$\int\Big[F(\varphi)\exp\left\{i\int[f_x(\varphi)-\varphi'(x)]\,\eta(x)\,dx\right\}\times$$
$$\times\det\left\{1+\frac{\delta\tilde f}{\delta\varphi}\right\}\prod_x d\varphi\Big]d\eta=F(\tilde\varphi), \qquad (5.21)$$

where $\tilde\varphi(\varphi')$ is the solution of the equation (5.20); $\det(1+\delta\tilde f/\delta\varphi)$

is, by definition,

$$\det\left\{1+\frac{\delta\tilde{f}}{\delta\varphi}\right\} \stackrel{\text{def}}{=} \exp\left\{\operatorname{Tr}\ln\left[1+\frac{\delta\tilde{f}(x)}{\delta\varphi(y)}\right]\right\} =$$
$$= \exp\left\{\int \frac{\delta\tilde{f}(x)}{\delta\varphi(y)}\bigg|_{x=y} dx + \frac{1}{2}\int dx\, dy\, \frac{\delta\tilde{f}(x)\,\delta\tilde{f}(y)}{\delta\varphi(y)\,\delta\varphi(x)} + \ldots\right\}. \tag{5.22}$$

According to the definition,

$$\int\left[\int \exp\left\{\frac{i}{2}\int \varphi(x)\,K(x-y)\,\varphi(y)\,dx\,dy + \right.\right.$$
$$\left. + i\int [f(\varphi)-\varphi'(x)]\,\eta(x)\,dx\right\} \det\left\{1+\frac{\delta\tilde{f}}{\delta\varphi}\right\}\prod_x d\varphi \prod_x d\eta =$$
$$= \int\left[\overrightarrow{\exp}\left\{i\int \eta(x)\tilde{f}\left(\frac{1}{i}\frac{\delta}{\delta\eta}\right)dx\right\}\det\left\{1+\frac{\delta\tilde{f}}{\delta\varphi}\left(\frac{1}{i}\frac{\delta}{\delta\eta}\right)\right\}\times\right.$$
$$\left.\times \exp\left\{-\frac{i}{2}\int \eta(x)\,K^{-1}(x-y)\,\eta(y)\,dx\,dy\right\}\right]\times$$
$$\times \exp\left\{-i\int [\varphi'(x)-c_0(x)]\,\eta(x)\,dx\right\}\prod_x d\eta. \tag{5.23}$$

The sign $\rightarrow$ ($\leftarrow$) over the exponential function shows that in the representation of the exponential function as a series all operators $\delta/\delta n$ must be placed to the right (left) of η. Integrating by parts, we transform the right-hand side into

$$\int \exp\left\{-\frac{i}{2}\int \eta(x)\,K^{-1}(x-y)\,\eta(y)\,dx\,dy\right\}\times$$
$$\times \det\left\{1+\frac{\delta\tilde{f}}{\delta\varphi}\left(\frac{1}{i}\frac{\delta}{\delta\eta}\right)\right\}\overleftarrow{\exp}\left\{i\int \tilde{f}\left(-\frac{1}{i}\frac{\delta}{\delta\eta}\right)\eta(x)\,dx\right\}\times$$
$$\times \exp\left\{-i\int [\varphi'(x)-c_0(x)]\,\eta(x)\,dx\right\}\prod_x d\eta. \tag{5.24}$$

Let us consider the functional

$$B(\varphi',\eta) =$$
$$= \det\left\{1+\frac{\delta\tilde{f}}{\delta\varphi}\left(\frac{1}{i}\frac{\delta}{\delta\eta}\right)\right\}\overleftarrow{\exp}\left\{i\int \tilde{f}\left(-\frac{1}{i}\frac{\delta}{\delta\eta}\right)\eta(x)\,dx\right\}\times$$

$$\times \exp\left\{-i\int[\varphi'(x)-c_0(x)]\,\eta(x)\,dx\right\}=$$
$$=\overleftarrow{\exp}\left\{-i\int\frac{\delta}{\delta\varphi'(x)}\tilde{f}(\varphi'-c_0)\,dx\right\}\times$$
$$\times \det\left\{1+\frac{\delta\tilde{f}}{\delta\varphi}(\varphi'-c_0)\right\}\exp\left\{-i\int[\varphi'(x)-c_0(x)]\,\eta(x)\,dx\right\}. \tag{5.25}$$

$B(\varphi', \eta)$ satisfies the equation

$$\delta B/\delta\eta(x)=i\left[c_0(x)-\varphi'(x)+\tilde{f}\left(i\frac{\delta}{\delta\eta}\right)\right]B \tag{5.26}$$

with the initial condition

$$B(\varphi',0)=A(\varphi')=\exp\left\{-\int\frac{\delta}{\delta\varphi'(x)}\tilde{f}(\varphi'-c_0)\,dx\right\}\times$$
$$\times\det\left\{1+\frac{\delta\tilde{f}}{\delta\varphi}(\varphi'-c_0)\right\}\cdot 1. \tag{5.27}$$

We shall look for the solution of the equation (5.26) in the form

$$B(\varphi',\eta)=A(\varphi')\exp\left\{-i\int\varphi(\varphi')\,\eta(x)\,dx\right\}. \tag{5.28}$$

Substituting (5.28) into (5.26), we obtain

$$\varphi(x)=\varphi'(x)-c_0(x)-\tilde{f}(\varphi). \tag{5.29}$$

Therefore the integral of interest is equal to

$$A(\varphi')\int\exp\left\{-\frac{i}{2}\int\eta(x)K^{-1}(x-y)\,\eta(y)\,dx\,dy\right\}\times$$
$$\times\exp\left\{-i\int\tilde{\varphi}(x)\,\eta(x)\,dx\right\}\prod_x d\eta=$$
$$=A(\varphi')\exp\left\{\frac{i}{2}\int\tilde{\varphi}(x)\,K(x-y)\,\tilde{\varphi}(y)\,dx\,dy\right\}. \tag{5.30}$$

It remains only for us to prove that $A(\varphi')=1$. The formula (5.27) can be rewritten as

$$A(\varphi') = \det\left[\overleftarrow{\exp}\left\{-\frac{\delta}{\delta\varphi'(x)}\tilde{f}(\varphi' - c_0)\right\}\times \right.$$
$$\left.\times\left\{1 + \frac{\delta\tilde{f}}{\delta\varphi}(\varphi' - c_0)\right\}\right]\cdot 1 = \det\left[\sum_n \frac{(-1)^n}{n!}\frac{\delta^n}{\delta\varphi'^n(x)}\times \right.$$
$$\left.\times\tilde{f}^n(\varphi' - c_0)\left\{1 + \frac{\delta\tilde{f}}{\delta\varphi'}(\varphi' - c_0)\right\}\right]\cdot 1. \quad (5.31)$$

Let us consider the nth term of the sum in square brackets. It is a binomial, the first term of which may be represented as

$$\frac{(-1)^n}{n!}\frac{\delta^n}{\delta\varphi'^n}\tilde{f}^n(\varphi') = \frac{(-1)^n}{(n-1)!}\frac{\delta^{n-1}}{\delta\varphi'^{n-1}}\left(\frac{\delta\tilde{f}}{\delta\varphi'}\tilde{f}^{n-1}(\varphi')\right). \quad (5.32)$$

On the other hand, the $(n-1)$th term of the sum in square brackets is represented by an analogous binomial, the second term of which is equal to

$$\frac{(-1)^{n-1}}{(n-1)!}\frac{\delta^{n-1}}{\delta\varphi'^{n-1}}\left(\frac{\delta\tilde{f}}{\delta\varphi'}\tilde{f}^{n-1}(\varphi')\right). \quad (5.33)$$

Thus, successive terms in square brackets cancel out, and the whole expression is equal to one.

4. Change of variables. Let

$$I = \int \exp\left\{\frac{i}{2}\int \varphi(x) K(x-y)\varphi(y)\,dx\,dy + \right.$$
$$\left. + i\int \varphi(x)\eta(x)\,dx\right\}\prod_x d\varphi. \quad (5.34)$$

By a change of variables

$$\varphi = f_x(\varphi'), \quad f_x(\varphi') = c_0(x) + \varphi'(x) + \tilde{f}(\varphi'), \quad (5.35)$$

I is reduced to the form

$$I = \int \exp\left\{\frac{i}{2}\int f_x(\varphi') K(x-y) f_y(\varphi')\,dx\,dy + \right.$$
$$\left. + i\int f_x(\varphi')\eta(x)\,dx\right\}\det\left\{1 + \frac{\delta\tilde{f}}{\delta\varphi'}\right\}\prod d\varphi'. \quad (5.36)$$

In order to prove this statement, it is sufficient to verify that the Fourier transforms of (5.34) and of (5.36) are equal to each other. The Fourier transform of (5.34) is

$$\exp\left\{\frac{i}{2}\int \psi(x) K(x-y) \psi(y) \, dx \, dy\right\}. \tag{5.37}$$

The Fourier transform of (5.36) equals

$$\begin{aligned}\tilde{I}(\psi) &= \int I(\eta) \exp\left\{-i\int \eta(x) \psi(x) \, dx\right\} \prod_x d\eta = \\ &= \exp\left\{\frac{i}{2}\int f_x K(x-y) f_y \, dx \, dy\right\} \det\left\{1+\frac{\delta \tilde{f}}{\delta \varphi'}\right\} \times \\ &\times \delta(\psi - f(\varphi')) \prod_x d\varphi' = \exp\left\{\frac{i}{2}\int \psi(x) K(x-y) \psi(y) \, dx \, dy\right\}.\end{aligned} \tag{5.38}$$

The statement is proved.

Our reasoning shows that all those properties of the Feynman integral which are used in practice in perturbation theory follow directly from the definition of the quasi-Gaussian integral and can be rigorously proved independently of the question of the existence of the Feynman-integral measure. Thus, in the framework of perturbation theory, the formalism of the path integral is a perfectly rigorous mathematical method, and results obtained with it do not need additional justification.

All these conclusions apply to the same extent to path integrals containing Fermi variables. In this case one must bear in mind the anticommutativity of variational derivatives, and in formulas for changing variables the corresponding determinant must be written in the denominator instead of the numerator. This characteristic feature of Gaussian integrals over Fermi variables has been already discussed above.

CHAPTER 3

QUANTIZATION OF THE YANG–MILLS FIELD

3.1. THE LAGRANGIAN OF THE YANG–MILLS FIELD AND THE SPECIFIC PROPERTIES OF ITS QUANTIZATION

In the previous chapter we used the path integral to formulate quantization rules for scalar and spinor fields, chosen as examples. At first sight it seems possible to quantize the Yang–Mills field in an analogous manner, considering each component of the field to be a scalar field. This, however, is not so. Gauge invariance introduces certain specific features into the quantization procedure. The spinor and scalar fields with which the Yang–Mills field interacts do not have any influence on these specific features. In the first three following sections we shall therefore restrict ourselves to the discussion of the Yang–Mills field in a vacuum.

We recall the notation introduced in the first chapter. Let Ω be a compact group of internal symmetry, T^a $(a = 1,\ldots,n)$ its orthonormalized generators in the adjoint representation, t^{abc} the corresponding structure constants, and

$$\mathcal{A}_\mu = A_\mu^a T^a \tag{1.1}$$

the Yang–Mills field. The gauge transformation is given by the matrix $\omega(x)$ with values in the adjoint representation of the group

$$\mathcal{A}_\mu(x) \to \mathcal{A}_\mu^\omega(x) = \omega(x)\,\mathcal{A}_\mu(x)\,\omega^{-1}(x) + \partial_\mu\omega(x)\omega^{-1}(x). \tag{1.2}$$

L. D. Faddeev and A. A. Slavnov. Gauge Fields: Introduction to Quantum Theory, ISBN 0-8053-9016-2.

Copyright © 1980 by The Benjamin/Cummings Publishing Company, Advanced Book Program. All rights reserved. No part of this publication may be reproduced, stored in a retrieval system, or transmitted, in any form or by any means, electronic, mechanical photocopying, recording, or otherwise, without the prior permission of the publisher.

The gauge-invariant Lagrangian has the form

$$\mathcal{L} = \frac{1}{8g^2} \operatorname{tr} \{\mathcal{F}_{\mu\nu} \mathcal{F}_{\mu\nu}\}, \tag{1.3}$$

where

$$\mathcal{F}_{\mu\nu} = \partial_\nu \mathcal{A}_\mu - \partial_\mu \mathcal{A}_\nu + [\mathcal{A}_\mu, \mathcal{A}_\nu]. \tag{1.4}$$

The equations of motion

$$\nabla_\mu \mathcal{F}_{\mu\nu} = \partial_\mu \mathcal{F}_{\mu\nu} - [\mathcal{A}_\mu, \mathcal{F}_{\mu\nu}] = 0 \tag{1.5}$$

are second-order equations with respect to $\mathcal{A}_\mu$ and are gauge-invariant: if $\mathcal{A}_\mu(x)$ is a solution of the equations of motion, then $\mathcal{A}_\mu^\omega(x)$ will also be a solution for any $\omega(x)$, where ω depends arbitrarily on x. This means that the equations of motion (1.5) are not independent. And, indeed, it is not difficult to verify that

$$\nabla_\nu \nabla_\mu \mathcal{F}_{\mu\nu} = 0. \tag{1.6}$$

To prove this, we represent $\nabla_\nu \nabla_\mu$ as

$$\nabla_\nu \nabla_\mu = \frac{1}{2} (\nabla_\nu \nabla_\mu + \nabla_\mu \nabla_\nu) + \frac{1}{2} (\nabla_\nu \nabla_\mu - \nabla_\mu \nabla_\nu). \tag{1.7}$$

We recall that for any matrix $\mathcal{B}(x)$ in the adjoint representation

$$(\nabla_\nu \nabla_\mu - \nabla_\mu \nabla_\nu) \mathcal{B}(x) = [\mathcal{F}_{\mu\nu}, \mathcal{B}]. \tag{1.8}$$

Since $\mathcal{F}_{\mu\nu}$ is antisymmetric with respect to μ and ν, we get

$$\nabla_\nu \nabla_\mu \mathcal{F}_{\mu\nu} = \frac{1}{2} [\mathcal{F}_{\mu\nu}, \mathcal{F}_{\mu\nu}] = 0. \tag{1.9}$$

The equality (1.9) is a special case of the second Noether theorem, which asserts that invariance of the Lagrangian with respect to transformations depending on an arbitrary function leads to the linear dependence of the equations of motion.

The abovementioned specific feature of the equations of motion manifests itself in their quantization. Indeed, some of the functions which parametrize the classical solution depend arbitrarily on time and do not obey the dynamics. While quantizing we must separate the true dynamical variables and the group parameters. We shall deal with this problem in the next section.

We shall now show why we cannot naïvely transfer the rules developed in the previous chapter for the construction of perturbation theory to the case of the Yang–Mills fields. According to the prescription of Chapter Two, in order to construct the perturbation theory for a given Lagrangian $\mathscr{L}$ one must represent it in the form

$$\mathscr{L} = \mathscr{L}_0 + \mathscr{L}_{\text{int}}, \tag{1.10}$$

where $\mathscr{L}_0$ is a form quadratic in the fields, and $\mathscr{L}_{\text{int}}$ contains higher-order forms in the fields. The monomials in $\mathscr{L}_{\text{int}}$ define the vertices with three or more legs, and $\mathscr{L}_0$ defines the propagators corresponding to internal lines. Namely, the propagator is the kernel of the integral operator that is inverse to the differential operator which defines the quadratic form $\mathscr{L}_0$.

Up to an inessential divergence, $\mathscr{L}_0$ for the Yang–Mills Lagrangian has the form (for the normalization $\mathscr{A}_\mu \to g\mathscr{A}_\mu$)

$$\begin{aligned}\mathscr{L}_0 = -\frac{1}{4}\left(\partial_\nu A^a_\mu - \partial_\mu A^a_\nu\right)\left(\partial_\nu A^a_\mu - \partial_\mu A^a_\nu\right) = \\ = -\frac{1}{2}\left[\partial_\mu A^a_\nu\, \partial_\mu A^a_\nu - \partial_\mu A^a_\mu\, \partial_\nu A^a_\nu\right].\end{aligned} \tag{1.11}$$

In the momentum representation the quadratic form $\mathscr{L}_0$ is given by the expression

$$K^{ab}_{\mu\nu}(k) = \delta^{ab}\left(g_{\mu\nu}k^2 - k_\mu k_\nu\right). \tag{1.12}$$

This operator has no inverse, and therefore the propagator is not defined. The reason is that, as has already been pointed out, not all components of the Yang–Mills field are independent dynamical variables. An analogous difficulty, as is known, is met in quantum electrodynamics. In this case one uses the Gupta–Bleuler formalism, which actually reduces to the following: as a photon propagator the function

$$D_{\mu\nu}(k^2) = -\frac{g_{\mu\nu}}{k^2 + i0}, \tag{1.13}$$

is chosen, and it is shown that the S-matrix constructed with its aid is unitary.

Generalizing this recipe, one could try to construct the Yang–Mills theory, using the propagator

$$D^{ab}_{\mu\nu} = -\frac{\delta^{ab} g_{\mu\nu}}{k^2 + i0} . \tag{1.14}$$

However, as was first shown by Feynman, such a construction of the perturbation theory is inadmissible. The S-matrix calculated with such a propagator is not unitary. Therefore it is necessary to revise the derivation of the perturbation-theory rules proceeding from the causal description of the classical dynamics for the Yang–Mills field, using for it the most convenient Hamiltonian formulation of this theory.

3.2. THE HAMILTONIAN FORMULATION OF THE YANG–MILLS FIELD AND ITS QUANTIZATION

In order to construct a consistent quantization procedure we must first find the true dynamical variables for the Yang–Mills field and verify that they change with time according to the laws of Hamiltonian dynamics. After this, we shall be able, in constructing the evolution operator, to use the path-integral formalism developed in the previous chapter. Let us consider in greater detail the structure of the Lagrangian of the Yang–Mills field. It is convenient to use the Lagrangian in the first-order formalism:

$$\mathcal{L} = \frac{1}{4} \operatorname{tr} \left\{ \left(\partial_\nu \mathcal{A}_\mu - \partial_\mu \mathcal{A}_\nu + g [\mathcal{A}_\mu, \mathcal{A}_\nu] - \frac{1}{2} \mathcal{F}_{\mu\nu} \right) \mathcal{F}_{\mu\nu} \right\}, \tag{2.1}$$

where $\mathcal{A}_\mu$ and $\mathcal{F}_{\mu\nu}$ are assumed to be independent variables. Obviously, this Lagrangian, and the equations of motion following from it, are equivalent to the Lagrangian (1.3).

In the three-dimensional notation ($\mu = 0, k$; $\nu = 0, l$; $k, l = 1, 2, 3$) we may rewrite the Lagrangian (up to a divergence) in the form

$$\mathcal{L} = -\frac{1}{2} \operatorname{tr} \left\{ \mathcal{E}_k \partial_0 \mathcal{A}_k - \frac{1}{2} (\mathcal{E}_k^2 + \mathcal{G}_k^2) + \mathcal{A}_0 \mathcal{C} \right\}, \tag{2.2}$$

where

$$\mathcal{E}_k = \mathcal{F}_{k0}, \quad \mathcal{G}_k = \frac{1}{2} \varepsilon^{ijk} \mathcal{F}_{ji}, \quad \mathcal{C} = \partial_k \mathcal{E}_k - g [\mathcal{A}_k, \mathcal{E}_k] \tag{2.3}$$

and we assume that $\mathcal{F}_{ik}$ is expressed in terms of $\mathcal{A}_i$ through the equations of motion, not including time derivatives:

$$\mathcal{F}_{ik} = \partial_k \mathcal{A}_i - \partial_i \mathcal{A}_k + g[\mathcal{A}_i, \mathcal{A}_k]. \tag{2.4}$$

This same Lagrangian can be written in the form

$$\mathcal{L} = E_k^a \partial_0 A_k^a - h(E_k, A_k) + A_0^a C^a, \quad h = \frac{1}{2}\{(E_k^a)^2 + (G_k^a)^2\}. \tag{2.5}$$

It is clear from the form of the Lagrangian (2.5) that the pairs (E_k^a, A_k^a) are canonical variables: h is the Hamiltonian, A_0^a is the Lagrangian multiplier, and C^a is the constraint on the canonical variables. By introducing Poisson brackets

$$\{E_k^a(x), A_l^b(y)\} = \delta_{kl}\delta^{ab}\delta(x-y), \tag{2.6}$$

it is easy to verify that

$$\{C^a(x), C^b(y)\} = gt^{abc}C^c(x)\delta(x-y) \tag{2.7}$$

and that

$$\left\{\int d^3x[(E_k^a)^2 + (G_k^a)^2], C^b(y)\right\} = 0. \tag{2.8}$$

This means that our system presents an example of the so-called generalized Hamilton dynamics. This concept was introduced by Dirac. Let us consider it, using as an example a system with n degrees of freedom. Let p_i and q_i be canonical variables, running through the phase space Γ^{2n}, and let the action have the form

$$A = \int\left[\sum_{i=1}^{n} p_i \dot{q}_i - h(p, q) - \sum \lambda^\alpha \varphi^\alpha(p, q)\right] dt;$$
$$\alpha = 1 \ldots m; \quad m < n. \tag{2.9}$$

Here the variables λ^α, additional to p and q, are called the Lagrangian multipliers; and φ^α are the constraints. Such an action defines a generalized Hamiltonian system if the conditions

$$\{h, \varphi^\alpha\} = c^{\alpha\beta}(p, q)\varphi^\beta; \quad \{\varphi^\alpha, \varphi^\beta\} = \sum_\gamma c^{\alpha\beta\gamma}(p, q)\varphi^\gamma \tag{2.10}$$

are fulfilled with certain coefficients $c^{\alpha\beta}$ and $c^{\alpha\beta\gamma}$, which in general depend on p, q. The generalized Hamiltonian system is equivalent to the usual Hamiltonian system Γ^* with $n - m$ degrees of freedom. The phase space $\Gamma^{*2(n-m)}$ of the latter system may be realized in the following manner. Let us consider m subsidiary conditions

$$\chi^m(p, q) = 0, \tag{2.11}$$

for which the requirements

$$\det|\{\varphi^\alpha, \chi^\beta\}| \neq 0, \tag{2.12}$$

$$\{\chi^\alpha, \chi^\beta\} = 0. \tag{2.13}$$

are satisfied. Then the subspace in Γ^{2n}

$$\chi^\alpha(p, q) = 0, \quad \varphi^\alpha(p, q) = 0 \tag{2.14}$$

is the space $\Gamma^{*2(n-m)}$ in question. The canonical variables p^*, q^* in $\Gamma^{*2(n-m)}$ may be found as follows. Owing to the condition (2.13), we can choose the canonical variables in Γ^{2n} in such a way that the χ^α will coincide with the first m variables of the coordinate type

$$q = (\chi^\alpha, q^*). \tag{2.15}$$

Let

$$p = (p^\alpha, p^*) \tag{2.16}$$

be the corresponding conjugate momenta. In these variables the condition (2.12) takes the form

$$\det\left|\frac{\partial\varphi^\alpha}{\partial p^\beta}\right| \neq 0, \tag{2.17}$$

so that the equations of the constraints

$$\varphi^\alpha(p, q) = 0 \tag{2.18}$$

may be solved for p^α. As a result, the subspace $p^{*2(n-m)}$ is given by the equations

$$\chi^\alpha \equiv q^\alpha = 0; \quad p^\alpha = p^\alpha(p^*, q^*) \tag{2.19}$$

and p^*, q^* are canonical. The Hamiltonian of this system is the function

$$h^*(p^*, q^*) = h(p, q)\,|_{\varphi=0,\ \chi=0}. \tag{2.20}$$

The equivalence of the systems Γ and Γ^* means the following. Let us consider the equations of motion for the system Γ:

$$\dot{p}_i + \frac{\partial h}{\partial q_i} + \lambda^\alpha \frac{\partial \varphi^\alpha}{\partial q_i} = 0, \quad \dot{q}_i - \frac{\partial h}{\partial p_i} - \lambda^\alpha \frac{\partial \varphi^\alpha}{\partial p_i} = 0, \tag{2.21}$$
$$\varphi^\alpha = 0.$$

The solutions of these equations contain arbitrary functions $\lambda^\alpha(t)$. Subsidiary conditions $\chi^\alpha(p, q) = 0$ remove this arbitrariness, by expressing $\lambda^\alpha(t)$ in terms of the canonical variables. As a result, the equations for the variables p^*, q^* are the only equations left. These equations coincide with the Hamiltonian equations for the system

$$\dot{q}^* = \frac{\partial h^*}{\partial p^*}; \quad \dot{p}^* = -\frac{\partial h^*}{\partial q^*}. \tag{2.22}$$

Indeed, consider the equations (2.19), (2.21) in terms of the coordinates (2.15), (2.16). The equations $\dot{q}^\alpha = 0$ lead to relationships which allow one to find λ^α:

$$\frac{\partial h}{\partial p_\alpha} + \lambda^\beta \frac{\partial \varphi_\beta}{\partial p_\alpha} = 0. \tag{2.23}$$

Let us now consider some one of the coordinates q^* and compare the equations for it which follow from (2.19), (2.21), and (2.22). They have the form

$$\dot{q}^* = \frac{\partial h}{\partial p^*} + \lambda^\alpha \frac{\partial \varphi_\alpha}{\partial p^*}, \tag{2.24}$$

$$\dot{q}^* = \frac{\partial h^*}{\partial p^*} = \frac{\partial h}{\partial p^*} + \frac{\partial h}{\partial p_\alpha} \frac{\partial p_\alpha}{\partial p^*} \tag{2.25}$$

respectively. The right-hand sides of these equations coincide if

$$\lambda^\alpha \frac{\partial \varphi_\alpha}{\partial p^*} = \frac{\partial h}{\partial p_\alpha} \frac{\partial p^\alpha}{\partial p^*}. \tag{2.26}$$

Using the equation (2.23), this condition may be rewritten as

$$\lambda^{\alpha}\left(\frac{\partial \varphi_{\alpha}}{\partial p^{*}}+\frac{\partial \varphi_{\alpha}}{\partial p_{\beta}}\frac{\partial p_{\beta}}{\partial p^{*}}\right)=0. \tag{2.27}$$

This equality holds owing to the conditions of constraint $\varphi_\alpha = 0$. The variables p^* are treated analogously. Thus, the statement is proved.

A change of choice of the subsidiary conditions is equivalent to a canonical transformation in the space $\Gamma^{*2(n-m)}$ and therefore does not influence the physics of the problem. For quantization of the system Γ the independent variables p^*, q^* may be used. Then the evolution operator is given by the path integral

$$\int \exp\left\{i\int [p^{*}\dot{q}^{*}-h(p^{*},\ q^{*})]\,dt\right\}\prod_{t}\frac{dp^{*}\,dq^{*}}{(2\pi)}, \tag{2.28}$$

where the initial and final values of the coordinates q^* are fixed. It is therefore desirable to be able to work directly in terms of the generalized system Γ. It is easy to verify that the path integral

$$\int \exp\left\{i\int [p_i\dot{q}_i-h(p,\ q)-\lambda^{\alpha}\varphi^{\alpha}(p,\ q)]\,dt\right\}\times \\ \times\prod_{t,\,\alpha}\delta(\chi^{\alpha})\prod_{t}\det|\{\varphi_{\alpha},\ \chi_{\beta}\}|\prod_{t}\frac{dp\,dq}{2\pi}\frac{d\lambda}{2\pi} \tag{2.29}$$

coincides with the integral (2.28). Indeed, integrating over λ, one can rewrite the formula (2.29) in the form

$$\int \exp\left\{i\int [p_i\dot{q}_i-h(p,\ q)]\,dt\right\}\times \\ \times\prod_{t,\,\alpha}\delta(\chi^{\alpha})\,\delta(\varphi^{\alpha})\prod_{t}\det|\{\varphi_{\alpha},\ \chi_{\beta}\}|\prod_{t}\frac{dp\,dq}{2\pi}. \tag{2.30}$$

In terms of the variables p^α, q^α, p^*, q^* the factor

$$\prod_{t}\delta(\varphi_{\alpha})\,\delta(\chi_{\alpha})\det|\{\varphi_{\alpha},\ \chi_{\beta}\}| \tag{2.31}$$

is rewritten as

$$\prod_{t}\delta(\varphi_{\alpha})\,\delta(q_{\alpha})\det\left|\frac{\partial\varphi_{\alpha}}{\partial p_{\beta}}\right|=\prod_{t}\delta(q_{\alpha})\,\delta[p_{\alpha}-p_{\alpha}(p^{*},\ q^{*})]. \tag{2.32}$$

As a result, the integral (2.29) is reduced by integration over p_α and q_α to (2.28).

Comparison of the formulas (2.5)–(2.8) and (2.9), (2.10) shows that the Yang–Mills field is indeed a generalized Hamilton system. We shall now apply the procedure just described to its quantization.

It is clear that in this case the gauge condition must play the role of the subsidiary condition. As such a condition we shall choose the relation

$$\partial_k \mathscr{A}_k = 0. \tag{2.33}$$

This condition is admissible. Indeed, it is obvious that

$$\{\partial_k A_k^a(x),\ \partial_i A_i^b(y)\} = 0. \tag{2.34}$$

Further

$$\{C^a(x),\ \partial_k A_k^b(y)\} = \partial_k [\partial_k \delta^{ab} - g t^{abc} A_k^c(x)] \delta(x - y). \tag{2.35}$$

The operator $M_c = \Delta\delta^{ab} - g t^{abc} A_k^c(x)\partial_k$ is reversible within the framework of perturbation theory. The inverse operator M_c^{-1} is an integral operator, the kernel $M_c^{-1}(x, y)$ of which is defined by the integral equation

$$M_C^{-1ab}(x, y) = \frac{1}{4\pi}\frac{\delta^{ab}}{|x - y|} + \\ + g\int \frac{dz}{4\pi} t^{acd} \frac{A_k^c(z)}{|x - z|} \partial_k M_C^{-1db}(z, y) \tag{2.36}$$

and may be calculated by iterations as a formal series in g. (Notice that for large fields A_k the operator M_c may have a nonzero eigenvalue, so that M_c^{-1} will cease to exist. This problem, however, is beyond the scope of perturbation theory, and we shall not discuss it here.)

The form of the subsidiary condition (2.33) suggests that in order to find the coordinates q^*, it is convenient to use the orthogonal expansion $\mathscr{A}_k$,

$$\mathscr{A}_k = \mathscr{A}_k^L + \mathscr{A}_k^T \tag{2.37}$$

in longitudinal and transverse components. Here

$$\mathscr{A}_K^L = \partial_k \mathscr{B}(x); \quad \mathscr{B}(x) = \frac{1}{4\pi}\int \frac{1}{|x - y|} \partial_k \mathscr{A}_k(y)\, dy, \tag{2.38}$$

where

$$\partial_k \mathcal{A}_k^T = 0. \tag{2.39}$$

It is clear that the transverse components $\mathcal{A}_k^T(x)$ play the role of q^*. The momenta which are their conjugates are the transverse components $\mathcal{E}_k^T(x)$. The equation for the constraints is the equation for the longitudinal part $\mathcal{E}_k^L(x)$. If one puts

$$\mathcal{E}_k^L(x) = \partial_k \mathcal{Q}(x), \tag{2.40}$$

then the equation for the constraint will be written as

$$\Delta \mathcal{Q} - g[\mathcal{A}_k, \partial_k \mathcal{Q}] - g[\mathcal{A}_k, \mathcal{E}_k^T] = 0, \tag{2.41}$$

where the operator M_c, already known to us, takes part. This equation allows us to express the longitudinal component $\mathcal{E}_k^L$ in terms of $\mathcal{E}_k^T$ and $\mathcal{A}_k^T$. After substituting the solution into the Hamiltonian $h(\mathcal{A}, \mathcal{E})$, we obtain the Hamiltonian $h^*(\mathcal{A}^T, \mathcal{E}^T)$ in the form of an infinite series in the constant g. The variables $\mathcal{A}^T$, $\mathcal{E}^T$ and the Hamiltonian h^* are the true Hamiltonian variables for the Yang–Mills field. The field configuration $\mathcal{A}^T$ at fixed time t is given by two functions of x. This means that the Yang–Mills field has two possible states of polarization.

We can now write down the S-matrix for the Yang–Mills field in terms of the path integral:

$$\begin{aligned} S = \lim_{\substack{t'' \to \infty \\ t' \to -\infty}} \int \exp\Big\{ i \int d^3k \, \frac{1}{2} \Big[\sum_{i=1,2} a_i^{*b}(\boldsymbol{k}, t'') a_i^b(\boldsymbol{k}, t'') + \\ + a_i^{*b}(\boldsymbol{k}, t') a_i^b(\boldsymbol{k}, t') \Big] + i \int_{t'}^{t''} dt \int d^3x \Big[\Big(-\frac{1}{4}\Big) \times \\ \times \operatorname{tr}\big[\mathcal{E}_l^T(\boldsymbol{x}, t) \dot{\mathcal{A}}_l^T(\boldsymbol{x}, t) - \dot{\mathcal{E}}_l^T(\boldsymbol{x}, t) \mathcal{A}_l^T(\boldsymbol{x}, t)\big] - \\ - h^*(\mathcal{E}^T, \mathcal{A}^T) \Big] \Big\} \prod \frac{d\overset{*}{a}_i(\boldsymbol{k}, t)\, da_i(\boldsymbol{k}, t)}{2\pi i}, \end{aligned} \tag{2.42}$$

where

$$\begin{aligned} A_l^{T,b}(\boldsymbol{x}, t) = \frac{1}{(2\pi)^{3/2}} \sum_{i=1,2} \int \big[e^{i\boldsymbol{k}\boldsymbol{x}} a_i^b(\boldsymbol{k}, t) u_l^i(\boldsymbol{k}) + \\ + e^{-i\boldsymbol{k}\boldsymbol{x}} a_i^{*b}(\boldsymbol{k}, t) u_l^i(-\boldsymbol{k}) \big] \frac{d^3k}{\sqrt{2\omega}}, \end{aligned}$$

$$E_l^{T,b}(\boldsymbol{x}, t) = \frac{i}{(2\pi)^{3/2}} \sum_{i=1,2} \int [-e^{ikx} a_i^b(\boldsymbol{k}, t) u_l^i(\boldsymbol{k}) + \\ + e^{-ikx} a_i^{*b}(\boldsymbol{k}, t) u_l^i(-\boldsymbol{k})] \frac{\sqrt{\omega}\, d^3k}{\sqrt{2}} \tag{2.43}$$

and $u_l^i(\boldsymbol{k})$, $i = 1, 2$, are two polarization vectors, which may be represented by two arbitrary orthonormalized vectors orthogonal to the vector $\boldsymbol{k}$. Here we assume that the asymptotic conditions

$$a_i^{*b}(\boldsymbol{k}, t'') \xrightarrow[t'' \to \infty]{} e^{i\omega t''} a_i^{*b}(\boldsymbol{k}); \quad a_i^b(\boldsymbol{k}, t') \xrightarrow[t' \to -\infty]{} e^{-i\omega t'} a_i^b(\boldsymbol{k}). \tag{2.44}$$

are fulfilled.

This formula is not very convenient for the construction of the diagram technique, since the Hamiltonian h^* is known only in the form of a series in the constant g, and besides, it generates vertices that are nonlocal with respect to the space coordinates. Of course, this is only a technical difficulty, but it strongly impedes practical calculations, in particular the construction of a renormalization procedure. This deficiency vanishes if we use the representation for the S-matrix in the form of an integral over all functions $\mathscr{A}_i(\boldsymbol{k}, t)$, $\mathscr{E}_i(\boldsymbol{k}, t)$:

$$S = \lim_{\substack{t'' \to \infty \\ t' \to -\infty}} \int \exp\Big\{ i \int d^3k \Big[\frac{1}{2} \sum_{i=1,2} a_i^{*b}(\boldsymbol{k}, t'') a_i^b(\boldsymbol{k}, t'') + \\ + a_i^{*b}(\boldsymbol{k}, t') a_i^b(\boldsymbol{k}, t') \Big] + i \int_{t'}^{t''} dt \int d^3x \times \\ \times \Big(-\frac{1}{4}\Big) \operatorname{tr} [\mathscr{E}_l(\boldsymbol{x}, t) \dot{\mathscr{A}}_l(\boldsymbol{x}, t) - \dot{\mathscr{E}}_l(\boldsymbol{x}, t) \mathscr{A}_l(\boldsymbol{x}, t) - \\ - \mathscr{E}_l^2(\boldsymbol{x}, t) - \mathscr{G}_l^2(x, t) + 2\mathscr{A}_0(\partial_l \mathscr{E}_l - g[\mathscr{A}_l, \mathscr{E}_l])] \Big\} \times \\ \times \prod_{x, t} \delta(\partial_l \mathscr{A}_l) \det M_C[\mathscr{A}] \prod_{x, t} d\mathscr{A}_l\, d\mathscr{E}_l\, d\mathscr{A}_0. \tag{2.45}$$

Here the boundary terms $a_i^*(\boldsymbol{k}, t'')$, $a_i(\boldsymbol{k}, t)$ are defined by the same formulas as before, that is, in terms of the transverse fields $\mathscr{A}_i^T$. We can now integrate over the momenta $\mathscr{E}_k$, taking into account the boundary conditions, as we did in Chapter Two in the case of a scalar field. As a result, we obtain for the normal symbol of the S-matrix the expression

$$S = N^{-1} \int \exp\left\{ i \int dx \left[\frac{1}{8} \operatorname{tr} \mathcal{F}_{\mu\nu} \mathcal{F}_{\mu\nu} \right] \right\} \times \\ \times \prod_x \delta(\partial_k \mathcal{A}_k) \prod_t \det M_C[\mathcal{A}] \prod_x d\mathcal{A}_\mu, \tag{2.46}$$

where integration is performed over all fields $\mathcal{A}_\mu(x)$, the asymptotic behavior of their three-dimensional parts

$$\mathcal{A}_l^T(x)_{t \to \pm\infty} \to \mathcal{A}^T_{l\, \substack{\text{out} \\ \text{in}}}(x);$$

$$A^{T,b}_{l, \substack{\text{in} \\ \text{out}}} = \frac{1}{(2\pi)^{3/2}} \sum_{i=1,2} \int \Big[a^b_{i, \substack{\text{in} \\ \text{out}}}(\boldsymbol{k})\, e^{ikx - i\omega t} u^i_l(\boldsymbol{k}) + \\ + a^{*b}_{i, \substack{\text{in} \\ \text{out}}}(\boldsymbol{k})^{-ikx + i\omega t} u^i_l(\boldsymbol{k}) \Big] \frac{d^3k}{\sqrt{2k_0}},$$

$$a^b_{i,\text{in}}(\boldsymbol{k}) = a^b_i(\boldsymbol{k}), \quad a^{*b}_{i,\text{out}}(\boldsymbol{k}) = a^{*b}_i(\boldsymbol{k}), \qquad i = 1, 2 \tag{2.47}$$

being fixed, and the definition of the quadratic form of the action being correspondingly supplemented. In the formula (2.46), as in the case of the scalar field, the Feynman functional exp($i \times$ action) is integrated. However, integration is not performed over all fields. The integration measure contains explicitly the δ-function of the gauge condition. This is a manifestation of the relativity principle, according to which it is necessary to integrate not over all fields, but only over classes of gauge-equivalent fields; the δ-function selects one representative from each class, and the determinant provides the correct normalization of the integration measure. The asymptotic conditions are also in accordance with the choice of the gauge condition.

Expansion of the integral (2.46) in a perturbation-theory series gives rise to the diagram technique. The propagator is defined by the Gaussian integral

$$I(J) = N^{-1} \int \exp\left\{ i \int dx \operatorname{tr} \left[\frac{1}{8} (\partial_\nu \mathcal{A}_\mu - \partial_\mu \mathcal{A}_\nu)^2 - \frac{1}{2} \mathcal{J}_\mu \mathcal{A}_\mu \right] \right\} \times \\ \times \prod_x \delta(\partial_k \mathcal{A}_k)\, d\mathcal{A}_\mu \tag{2.48}$$

with the Feynman boundary conditions for $\mathcal{A}_k^T$. This integral is equal to

$$\exp\left\{\frac{i}{2}\int J_\mu^a(x)\,D_{\mu\nu}^c(x-y)\,J_\nu^a(y)\,dx\,dy\right\}, \tag{2.49}$$

where $D_{\mu\nu}^c$ is the propagator sought:

$$\begin{aligned}
D_{ml}^c(x) &= -\frac{1}{(2\pi)^4}\int e^{-ikx}\left(\delta^{ml}-\frac{k_m k_l}{|\boldsymbol{k}|^2}\right)(k^2+i0)^{-1}\,dk,\\
D_{m0}^c(x) &= D_{0m}^c(x)=0;\\
D_{00}^c(x) &= -\frac{1}{(2\pi)^4}\int e^{-ikx}\frac{1}{|\boldsymbol{k}|^2}\,dk.
\end{aligned} \tag{2.50}$$

For a proof we shall use the integral representation of the δ-function (2.5.18). Then $I(J)$ is given by the Gaussian integral

$$\begin{aligned}
I(J) = N^{-1}\int \exp\Big\{i\int dx\Big[-\frac{1}{4}(\partial_\nu A_\mu^l-\partial_\mu A_\nu^l)^2 &+ \\
+ J_\mu^l A_\mu^l + \lambda^l\partial_k A_k^l\Big]\Big\}&\prod_x dA_\mu\,d\lambda,
\end{aligned} \tag{2.51}$$

for the calculation of which it is necessary to find the extremum of the exponent. The equations

$$\begin{aligned}
&\partial_\nu(\partial_\nu A_k^l-\partial_k A_\nu^l)+J_k^l+\partial_k\lambda^l=0,\\
&\partial_\nu(\partial_\nu A_0^l-\partial_0 A_\nu^l)+J_0^l=0;\quad \partial_k A_k^l=0
\end{aligned} \tag{2.52}$$

are rewritten as

$$\begin{aligned}
&\Box A_k^l+(\partial_k\lambda^l+\partial_0\partial_k A_0^l)+J_k^l=0,\\
&\triangle A_0^l-J_0^l=0;\quad \partial_k A_k^l=0
\end{aligned} \tag{2.53}$$

and have unique solutions under the above-formulated boundary conditions. It may be assumed also that the source J satisfies the transversality condition

$$\partial_k J_k^l=0. \tag{2.54}$$

As a result, the solution is given by the formula

$$A_\mu^l(x)=\int D_{\mu\nu}^c(x-y)\,J_\nu^l(y)\,dy, \tag{2.55}$$

where $D_{\mu\nu}^c(x)$ is the Coulomb propagator just introduced.

The explicit expression for $D^c_{\mu\nu}$ shows that only three-dimensionally transverse components $\mathcal{A}_\mu$ propagate in time, in agreement with our boundary conditions.

A defect of the diagram technique in the Coulomb gauge is the lack of explicit relativistic invariance. In the next section we shall show that in the integral (2.46) defining the S-matrix, one can pass to the manifestly covariant gauge.

To conclude this section we shall give a description of the alternative Hamiltonian formulation of the Yang–Mills theory, using the gauge condition $A_0 = 0$. This gauge is an improvement on the Coulomb one in that it is also admissible beyond the scope of perturbation theory. Let us show that in each class of gauge-equivalent fields there exists a field satisfying the condition

$$\mathcal{A}_0 = 0. \tag{2.56}$$

For this we point out that the equation

$$\frac{\partial}{\partial t}\,\omega(\boldsymbol{x},\, t) = -\,\omega(\boldsymbol{x},\, t)\,\mathcal{A}_0(\boldsymbol{x},\, t) \tag{2.57}$$

allows a solution of the form

$$\omega_0(\boldsymbol{x},\, t) = T\exp\left\{-\int_{-\infty}^{t} \mathcal{A}_0(\boldsymbol{x},\, s)\, ds\right\}, \tag{2.58}$$

where the symbol T signifies that the exponential is to be ordered in time. From the equation (2.57) it follows that

$$\mathcal{A}_\mu^{\omega_0} = \omega_0 \mathcal{A}_\mu \omega_0^{-1} + \partial_\mu \omega_0 \omega_0^{-1} \tag{2.59}$$

satisfies the condition

$$\mathcal{A}_0^{\omega_0} = 0. \tag{2.60}$$

In addition to the matrix $\omega_0(x)$, an analogous property is possessed by the matrices $\omega(x)$ of the form

$$\omega(x) = \omega(\boldsymbol{x})\,\omega_0(x), \tag{2.61}$$

where $\omega(\boldsymbol{x})$ is an arbitrary matrix from Ω, depending only on space coordinates. Thus, the Hamilton gauge does not completely abolish gauge arbitrariness in the definition of the Yang–Mills field, but reduces the gauge group to the group of matrices $\omega(\boldsymbol{x})$.

We shall now show that the equations of motion in the gauge $\mathscr{A}_0 = 0$ are actually Hamiltonian. For this it is convenient to use the equations of motion, formulated as first-order equations, following from the Lagrangian (2.1):

$$\partial_\nu \mathscr{A}_\mu - \partial_\mu \mathscr{A}_\nu + g[\mathscr{A}_\mu, \mathscr{A}_\nu] - \mathscr{F}_{\mu\nu} = 0, \qquad (2.62)$$
$$\partial_\mu \mathscr{F}_{\mu\nu} - g[\mathscr{A}_\mu, \mathscr{F}_{\mu\nu}] = 0.$$

Consider these equations in the three-dimensional formulation. In the notation $\mu = (0, k)$, $\nu = (0, l)$, etc., the 10 equations (2.62) are rewritten as

$$\begin{aligned}
&\partial_0 \mathscr{F}_{0k} = \partial_l \mathscr{F}_{lk} - g[\mathscr{A}_l, \mathscr{F}_{lk}],\\
&\partial_0 \mathscr{A}_k = \mathscr{F}_{k0},\\
&\mathscr{F}_{ik} = \partial_k \mathscr{A}_i - \partial_i \mathscr{A}_k + g[\mathscr{A}_i, \mathscr{A}_k],\\
&\mathscr{C}(x) = \partial_k \mathscr{F}_{0k} - g[\mathscr{A}_k, \mathscr{F}_{0k}] = 0.
\end{aligned} \qquad (2.63)$$

Eliminating the variables $\mathscr{F}_{ik}$ with the help of the equations of motion, we see that the system of equations (2.63) has an explicitly Hamiltonian form

$$\partial_0 E_k^a(\boldsymbol{x}, t) = -\frac{\delta H}{\delta A_k^a(\boldsymbol{x}, t)} = \{H, E_k^a(\boldsymbol{x}, t)\}, \qquad (2.64)$$
$$\partial_0 A_k^a(\boldsymbol{x}, t) = \frac{\delta H}{\delta E_k^a(\boldsymbol{x}, t)} = \{H, A_k^a(\boldsymbol{x}, t)\},$$
$$H = \int h\, d^3x,$$

where the above-introduced notation for E_k, h, and the Poisson bracket is used. The last equation of (2.63)

$$\mathscr{C}(\boldsymbol{x}, t) = 0 \qquad (2.65)$$

represents an equation of constraint. As we have already seen, the Poisson bracket $\{H, \mathscr{C}(\boldsymbol{x}, t)\}$ vanishes,

$$\{H, \mathscr{C}(\boldsymbol{x}, t)\} = \partial_0 \mathscr{C}(\boldsymbol{x}, t) = 0, \qquad (2.66)$$

so that $\mathscr{C}(\boldsymbol{x}, t)$ generates an infinite set of integrals of motion.

We shall show that $\mathscr{C}(\boldsymbol{x})$ are generators of infinitesimal gauge transformations, remaining after the imposition of the gauge condition $\mathscr{A}_0 = 0$. To do this, we associate with an arbitrary matrix $\alpha(\boldsymbol{x})$ in the adjoint representation $\Omega(x)$ the quantity

$$C(\alpha) = -\frac{1}{2}\operatorname{tr}\left\{\int \mathscr{C}(x)\,\alpha(x)\,d^3x\right\}. \tag{2.67}$$

The commutation relations (2.7) in this notation are rewritten as

$$\{C(\alpha),\ C(\beta)\} = gC([\alpha,\ \beta]). \tag{2.68}$$

This shows that $C(\alpha)$ defines the Lie-algebra representation of the group of gauge transformations, consisting of matrices $\alpha(x)$. The action of this representation on the variables $\mathscr{A}_l(x)$ and $\mathscr{E}_l(x)$ is given by the formula

$$\begin{aligned} \delta\mathscr{A}_l &= \{C(\alpha),\ \mathscr{A}_l(x)\} = \partial_l\alpha(\boldsymbol{x}) - g[\mathscr{A}_l(x),\ \alpha(\boldsymbol{x})], \\ \delta\mathscr{E}_k &= \{C(\alpha),\ \mathscr{E}_k(x)\} = -g[\mathscr{E}_k(x),\ \alpha(\boldsymbol{x})]. \end{aligned} \tag{2.69}$$

Thus, indeed, the $C(\alpha)$ are generators of gauge transformations, remaining in the Hamilton gauge.

In accordance with the relativity principle the observables $O(\mathscr{A}_k,\ \mathscr{E}_k)$ are gauge-invariant and therefore must commute with $C(\alpha)$. This condition is a system of first-order differential equations, for which the relation (2.68) plays the role of the integrability condition and expresses one of the six functions $\mathscr{A}_k$, $\mathscr{E}_k$, upon which O depends, in terms of all the others. Together with the conditions of constraint (2.65), this reduces the number of independent functions to four, in agreement with the calculated number of degrees of freedom in the Coulomb gauge.

Let us see how this classical picture is transferred to the quantum case. In the operator formulation the Hamiltonian, the constraint C, and the observables O become operators, which satisfy the relations

$$\begin{aligned} &\frac{1}{i}[\boldsymbol{C}(\alpha),\ \boldsymbol{C}(\beta)] = g\boldsymbol{C}([\alpha,\ \beta]), \\ &[H,\ \boldsymbol{C}(\alpha)] = 0; \quad [O,\ \boldsymbol{C}(\alpha)] = 0. \end{aligned} \tag{2.70}$$

We cannot directly equate the operator $\boldsymbol{C}$ to zero, although the formulas (2.70) show that there exists a subspace, formed by the vectors ψ, satisfying the equation

$$C\psi = 0, \tag{2.71}$$

and that this subspace is invariant with respect to the operators corresponding to the observables. The condition (2.71) replaces the classical equation $\boldsymbol{C} = 0$, and the constructed subspace is the true space of states of our physical system.

For a description of the dynamics it is not necessary to work in the physical subspace. It is simpler to consider the operator $\exp\{-iHt\}$ in the whole space and to impose the condition (2.71) only on the states between which the matrix elements are calculated. Since H and $\boldsymbol{C}$ commute, such a procedure is consistent. Passing to the S-matrix, note that the condition (2.71) is simplified for asymptotic states. It may be shown that

$$\lim_{|t|\to\infty} e^{iH_0t}\boldsymbol{C}(\alpha)\, e^{-iH_0t} = \boldsymbol{C}_0(\alpha) + O(1), \tag{2.72}$$

where

$$\boldsymbol{C}_0(\alpha) = -\frac{1}{2}\int \operatorname{tr} \partial_k \mathscr{E}_k(\boldsymbol{x}, t)\, \alpha(x)\, d^3x \tag{2.73}$$

is a generator of linearized gauge transformations:

$$\delta\mathscr{E}_k = 0; \quad \delta\mathscr{A}_k = \partial_k \alpha(x), \tag{2.74}$$

and H_0 is the free operator for the energy:

$$H_0 = -\frac{1}{4} \operatorname{tr} \int \left(\mathscr{E}_k^2 + (\partial_k\mathscr{A}_l - \partial_l\mathscr{A}_k)^2\right) d^3x. \tag{2.75}$$

Indeed, the difference between $\boldsymbol{C}(\alpha)$ and $\boldsymbol{C}_0(\alpha)$, having the form

$$\boldsymbol{C}_1(\alpha) = \operatorname{tr} \int [\mathscr{E}_k, \mathscr{A}_k]\, \alpha\, d^3x \tag{2.76}$$

is quadratic in the fields $\mathscr{A}$ and $\mathscr{E}$, and therefore the coefficient functions of the operator $\exp\{-iH_0t\}\ \boldsymbol{C}_1(\alpha)\ \exp\{iH_0t\}$ decrease as $|t| \to \infty$. The operator $\boldsymbol{C}_0$ commutes with H_0 and the S-matrix. This is shown by the following formal computation:

$$SC_0 = \lim_{\substack{t''\to+\infty \\ t'\to-\infty}} e^{iH_0t''}e^{-iH(t''-t')}e^{-iH_0t'}C_0 =$$
$$= \lim_{\substack{t''\to+\infty \\ t'\to-\infty}} e^{iH_0t''}e^{-iH(t''-t')}, Ce^{-iH_0t'} =$$
$$= \lim_{\substack{t''\to\infty \\ t'\to-\infty}} e^{iH_0t''}Ce^{-iH(t''-t')}e^{-iH_0t'} =$$
$$= \lim_{\substack{t''\to\infty \\ t'\to-\infty}} C_0e^{iH_0t''}e^{-iH(t''-t')}e^{-iH_0t'} = C_0S. \quad (2.77)$$

The state vectors $\psi(a_l^*)$, satisfying the asymptotic condition

$$C_0\psi = 0, \quad (2.78)$$

are given by the formula

$$\psi(a^*) = \exp\left\{\frac{1}{2}\int a^{*L}(\boldsymbol{k})\, a^L(\boldsymbol{k})\, d^3k\right\}\tilde{\psi}(a^{*T}), \quad (2.79)$$

where a^L and a^T stand for the components $a(\boldsymbol{k})$ parallel and orthogonal to the vector $\boldsymbol{k}$, respectively. As is seen, the vectors $\psi(a^*)$ actually depend only on two polarizations. According to the proof, presented above, this subspace is invariant for the S-matrix.

In terms of the path integral the reasoning given above is reformulated in the following manner. The S-matrix in the whole space is defined by the path integral

$$S = \lim_{\substack{t''\to\infty \\ t'\to-\infty}} \int \exp\Big\{\int d^3k\, \frac{1}{2}\big[a_l^{*b}(\boldsymbol{k}, t'')\, a_l^b(\boldsymbol{k}, t'') +$$
$$+ a_l^{*b}(\boldsymbol{k}, t')\, a_l^b(\boldsymbol{k}, t')\big] +$$
$$+ i\int_{t''}^{t'} dt \int d^3x \left(-\frac{1}{4}\right) \mathrm{tr}\big[\mathscr{E}_l(\boldsymbol{x}, t)\dot{\mathscr{A}}_l(\boldsymbol{x}, t) - \mathscr{E}_l(\boldsymbol{x}, t)\mathscr{A}_l(\boldsymbol{x}, t) -$$
$$- \mathscr{E}_l^2(\boldsymbol{x}, t) - \mathscr{G}_l^2(\boldsymbol{x}, t)\big]\Big\} \prod_{\boldsymbol{x},\, t} d\mathscr{A}_l\, d\mathscr{E}_l. \quad (2.80)$$

Integrating over $\mathscr{E}$, we obtain

$$S = \int_{\mathscr{A}_k \to \mathscr{A}_{k,\,\mathrm{out}}^{\mathrm{in}}} \exp\left\{i\int dx\left[-\frac{\mathrm{tr}}{4}\left(\dot{\mathscr{A}}_l^2 - \mathscr{G}_l^2\right)\right]\right\} \prod_x d\mathscr{A}_k, \quad (2.81)$$

where, in contrast with the analogous formula in the Coulomb gauge, boundary conditions are imposed on all the three components of $\mathscr{A}_k$.

The action in the exponential may be reduced to a relativistic-invariant form by introducing formal integration over $\mathscr{A}_0$. Then we obtain

$$S = \int_{\mathscr{A}_k \to \mathscr{A}_{k,\,\mathrm{out}}^{\mathrm{in}}} \exp\left\{i \int dx \left[\frac{1}{8} \operatorname{tr} \mathscr{F}_{\mu\nu}\mathscr{F}_{\mu\nu}\right]\right\} \prod_x \delta(\mathscr{A}_0)\, d\mathscr{A}_\mu. \quad (2.82)$$

This integral also allows interpretation in the spirit of the relativity principle. It is an integral over the classes of gauge-invariant fields under another gauge condition, defining the choice of representatives. The integration measure is simpler than in the Coulomb case and does not contain a determinant.

The propagator for the diagram technique has the form

$$D_{lm}^H = -\frac{1}{(2\pi)^4} \int e^{-ikx} \frac{1}{k_0^2} [\delta_{lm} + (\boldsymbol{k}^2 \delta_{lm} - k_l k_m)(\boldsymbol{k}^2 + i0)^{-1}]\, dk. \quad (2.83)$$

For its calculation it is necessary to solve the equations

$$\partial_\mu (\partial_\mu \mathscr{A}_k - \partial_k \mathscr{A}_\mu) + \mathscr{J}_k = 0, \quad \mathscr{A}_0 = 0, \quad (2.84)$$

which under our boundary conditions have the solution

$$\mathscr{A}_k = \mathscr{A}_k^{(0)} + \int D_{kl}^H (x - y)\, \mathscr{J}_l(y)\, dy. \quad (2.85)$$

This formula shows that, as in the Coulomb case, only the three-dimensionally transverse components of A_k^T propagate in time.

3.3 COVARIANT QUANTIZATION RULES AND THE FEYNMAN DIAGRAM TECHNIQUE

As was already pointed out, the expression for the S-matrix obtained in the previous section is not manifestly covariant. This is inconvenient for performing calculations within the framework of perturbation theory especially for renormalization procedures. The path-integral method allows us to get rid of this defect. The relativity principle suggests that for this it is necessary to pass to a relativisitic-invariant parametrization of the classes of gauge-equivalent fields, that is, to choose a relativistic-invariant gauge. The most simple relativitistic-invariant gauge condition is the Lorentz condition

$$\partial_\mu \mathcal{A}_\mu = 0. \tag{3.1}$$

We shall show how to pass to the Lorentz gauge starting with the already known expression for the S-matrix in the Coulomb gauge (2.46). From a geometrical point of view we must transfer the measure, defined on the surface $\Phi_C \equiv \partial_k \mathcal{A}_k = 0$, to the surface $\Phi_L \equiv \partial_\mu \mathcal{A}_\mu = 0$ along the trajectories of the gauge group. Formally this may be achieved in the following way. We introduce the functional $\triangle_L(A)$, proceeding from the condition

$$\triangle_L(\mathcal{A}) \int \prod_x \delta\left(\partial_\mu \mathcal{A}_\mu^\omega\right) d\omega = 1, \tag{3.2}$$

where integration is performed over the measure $\prod_x d\omega(x)$ and $d\omega$ is the invariant measure on the group Ω:

$$d(\omega\omega^0) = d(\omega^0\omega) = d\omega. \tag{3.3}$$

The functional $\triangle_L(\mathcal{A})$ is obviously gauge-invariant:

$$\triangle_L(\mathcal{A}^\omega) = \triangle_L(\mathcal{A}), \tag{3.4}$$

which follows directly from the invariance of the integration measure.

Using the relation (3.2), it is possible to rewrite the expression for the S-matrix (2.46) as

$$S = N^{-1} \int \exp\left\{ i \int dx \left[\frac{1}{8} \operatorname{tr} \mathcal{F}_{\mu\nu} \mathcal{F}_{\mu\nu} \right] \right\} \prod_x \delta(\partial_k \mathcal{A}_k) \times$$
$$\times \prod_t \det M_C(\mathcal{A}) \triangle_L(\mathcal{A}) \prod_x \delta\left(\partial_\mu \mathcal{A}_\mu^\omega\right) d\omega \, d\mathcal{A}. \tag{3.5}$$

Now note that the functional $\prod_t \det M_c(\mathcal{A})$ coincides with the gauge-invariant functional $\triangle_c(\mathcal{A})$ on the surface $\Phi_c = \partial_k \mathcal{A}_k = 0$, where $\triangle_c(\mathcal{A})$ is introduced analogously to $\triangle_L(\mathcal{A})$:

$$\triangle_C(\mathcal{A}) \int \delta\left(\partial_k \mathcal{A}_k^\omega\right) d\omega = 1. \tag{3.6}$$

Indeed, if $\mathcal{A}_k$ satisfies the condition $\partial_k \mathcal{A}_k = 0$, then $\omega = 1$ is obviously the root of the argument of the δ-function (the only one within the framework of perturbation theory). Therefore in the integral (3.6) it is sufficient to integrate only in the vicinity of the unit element. For $\omega(x) \approx 1 + u(x)$ we have

$$\partial_k \mathcal{A}_k^{\omega} = \Delta u - g[\mathcal{A}_k(x), \partial_k u(x)] = M_C u(x) \qquad (3.7)$$

and

$$\prod_x d\omega(x) = \prod_x du(x). \qquad (3.8)$$

Thus the integral is calculated explicitly, and we obtain that

$$\Delta_C(\mathcal{A})\,|_{\partial_k \mathcal{A}_k = 0} = \prod_t \det M_C(\mathcal{A}). \qquad (3.9)$$

Let us go back to our integral (3.5), in which, as just shown, we can put

$$\prod_x \delta(\partial_k \mathcal{A}_k) \prod_t \det M_C = \prod_x \delta(\partial_k \mathcal{A}_k)\, \Delta_C. \qquad (3.10)$$

We perform a change of variables

$$\mathcal{A}_\mu \to \mathcal{A}_\mu^{\omega^{-1}}, \qquad (3.11)$$

the Jacobian of which is obviously equal to unity.

Owing to the invariance of the action and factors Δ_L, Δ_c, the integral (3.5) can be rewritten in the form

$$S = N^{-1} \int \exp\left\{ i \int dx \left[\frac{1}{8} \operatorname{tr} \mathcal{F}_{\mu\nu} \mathcal{F}_{\mu\nu} \right] \right\} \prod_x \delta(\partial_\mu \mathcal{A}_\mu)\, \Delta_L(\mathcal{A}) \times$$
$$\times\, \delta\left(\partial_k \mathcal{A}_k^{\omega^{-1}}\right) \Delta_C(\mathcal{A})\, d\omega\, d\mathcal{A}. \qquad (3.12)$$

Substituting $\mathcal{A}^{\omega}$ in the integral over $d\omega$ for $\mathcal{A}^{\omega^{-1}}$ and using the formula (3.6), we see that the last two factors in the integrand of (3.12) may be dropped. As a result, we obtain the expression for the S-matrix in the Lorentz gauge:

$$S =$$
$$= N^{-1} \int \exp\left\{ i \int dx \left[\frac{1}{8} \operatorname{tr} \mathcal{F}_{\mu\nu} \mathcal{F}_{\mu\nu} \right] \prod_x \Delta_L(\mathcal{A})\, \delta(\partial_\mu \mathcal{A}_\mu)\, d\mathcal{A}. \right. \qquad (3.13)$$

Reasoning entirely analogous to the one which led to (3.9) shows that on the surface $\Phi_L \equiv \partial_\mu \mathscr{A}_\mu(x) = 0$ the functional $\triangle_L$ is equal to

$$\triangle_L(\mathscr{A})|_{\partial_\mu \mathscr{A}_\mu = 0} = \det M_L, \tag{3.14}$$

where the operator M_L is defined by the formula

$$M_L \alpha(x) = \Box \alpha - g\partial_\mu [\mathscr{A}_\mu, \alpha] = \Box \alpha + W(\mathscr{A})\alpha. \tag{3.15}$$

We recall that we have already encountered the determinants $\det M_C$ and $\det M_L$ which appear here in the first chapter, while formulating the admissibility of the gauge condition.

We have not yet discussed the influence of a change of variables on the asymptotic conditions in the integral (2.46). Therefore the formula (3.13) for the S-matrix is as yet somewhat formal. In particular, our reasoning does not make it clear what meaning is to be attributed to the determinant of M_L. The point is that for a consistent definition of the operator M_L in the whole space of variables x we need boundary conditions as $t \to \pm\infty$. In another manner this problem may be formulated as follows. For defining the determinant it is natural to use the formula

$$\det M_L = \exp\{\operatorname{Tr} \ln M_L\} = \\ = \exp\{\operatorname{Tr} \ln \Box + \operatorname{Tr} \ln (1 + \Box^{-1} W(\mathscr{A}))\}. \tag{3.16}$$

Here the symbol Tr stands for the operation of taking the trace, including also integration over the coordinates.

The first factor is an insignificant constant which only changes the normalization constant N. The second factor generates an additional term in the action which has the form

$$\operatorname{Tr} \ln (1 + \Box^{-1} W(\mathscr{A})) = \sum_n \frac{(-1)^{n+1}}{n} \operatorname{Tr} (\Box^{-1} W)^n = \\ = -\frac{g^2}{2} \int dx_1\, dx_2 \operatorname{tr} \{\mathscr{A}_{\mu_1}(x_1) \mathscr{A}_{\mu_2}(x_2)\} \partial_{\mu_1} D(x_1 - x_2) \times \\ \times \partial_{\mu_2} D(x_2 - x_1) - \ldots + (-1)^{n+1} \frac{g^n}{n} \int dx_1 \ldots dx_n \times \\ \times \operatorname{tr} \{\mathscr{A}_{\mu_1}(x_1) \ldots \mathscr{A}_{\mu_n}(x_n)\} \times \\ \times \partial_{\mu_1} D(x_1 - x_2) \ldots \partial_{\mu_n} D(x_n - x_1) - \ldots, \tag{3.17}$$

where $D(x)$ is the Green function of the d'Alembertian operator. This Green function is not defined uniquely, and the question arises which boundary conditions must be imposed for its unique definition. In practice, it is a question of how to get round the pole in the integral

$$D = -\frac{1}{(2\pi)^4}\int \frac{e^{-ikx}}{k^2}\,dk, \tag{3.18}$$

defining the Green function. An analogous problem arises in defining the Green function $D^L_{\mu\nu}(x-y)$ corresponding to the quadratic form in the Lorentz gauge. The formal answer, obtained by inverting this quadratic form, is

$$D^L_{\mu\nu}(x-y) = -\frac{1}{(2\pi)^4}\int e^{-ik(x-y)}\left\{g^{\mu\nu} - \frac{k_\mu k_\nu}{k^2}\right\}\frac{1}{k^2}\,dk. \tag{3.19}$$

In this formula it is also necessary to clarify in which sense one gets round the poles of the integrand.

For an answer to the question on the boundary conditions it is necessary to perform a transformation of the integral (2.46) into the integral (3.13) before passing to the time limit $t'' \to \infty$, $t' \to -\infty$. We recall that in the Coulomb gauge, besides the boundary conditions on the three-dimensional components of the potential $\mathcal{A}^T_k$, there exists the condition

$$\partial_k \mathcal{A}_k = 0, \tag{3.20}$$

which is satisfied in the whole interval $t' \leq t \leq t''$, including $t = t''$, $t = t'$. The change of variables

$$\mathcal{A}_\mu \to \mathcal{A}_\mu^{\omega^{-1}} = \omega^{-1}\mathcal{A}_\mu\omega + \partial_\mu\omega^{-1}\omega \tag{3.21}$$

should not violate this condition.

Hence follows the restriction on ω:

$$\omega(\boldsymbol{x}, t'') = 1, \quad \omega(\boldsymbol{x}, t') = 1, \tag{3.22}$$

which provides for the disappearance of the space derivatives $\partial_k\omega$. Note that the time derivative $\partial_0\omega$ need not necessarily disappear at $t = t'$ and that $t = t''$, since in the integral (2.46) no conditions are imposed on $\mathcal{A}_0$ at $t = t'$ and $t = t''$. Such a transformation also does not alter the boundary values of the transverse components of $\mathcal{A}^T_k$ as $t' \to -\infty$, $t'' \to +\infty$, since in this limit the transformation (3.21) is linearized and is reduced to the substitution

$$\mathscr{A}_\mu \to \mathscr{A}_\mu - \partial_\mu \alpha, \tag{3.23}$$

where

$$\omega = \exp\{\alpha\}. \tag{3.24}$$

Thus, the formal definition of the operator M_L, given by the formula (3.15), must be supplemented by the boundary conditions

$$\alpha(\boldsymbol{x}, t'') = \alpha(\boldsymbol{x}, t') = 0. \tag{3.25}$$

The Green function D, appearing in the expansion of the determinant in the perturbation-theory series, is the Green function of the d'Alembertian operator with the same boundary conditions. Such a function has the form

$$D_1(x, y) = \\ = \frac{1}{(2\pi)^3} \int e^{ik(x-y)} \frac{\sin[|\boldsymbol{k}|(x_0 - t')] \cdot \sin[|\boldsymbol{k}|(y_0 - t'')]}{|\boldsymbol{k}| \sin[|\boldsymbol{k}|(t'' - t)]} d^3k, \\ x_0 \leqslant y_0 ; \tag{3.26}$$

at $x_0 \geq y$, $D_1(x, y)$ is determined by the symmetry condition

$$D_1(x, y) = D_1(y, x).$$

With such a definition of the operator M_L, its determinant is positive in the framework of perturbation theory, and this justifies its use in the formula (3.14) instead of $|\det M_L|$.

The problem of getting round the poles in the Green function $D^L_{\mu\nu}$ is solved in an analogous manner. For its definition at finite t', t'' we must solve the equations

$$\Box \mathscr{A}_\mu = \mathscr{I}_\mu, \quad \partial_\mu \mathscr{A}_\mu = 0, \tag{3.27}$$

where $\mathscr{I}_\mu$ satisfies the compatibility condition

$$\partial_\mu \mathscr{I}_\mu = 0. \tag{3.28}$$

The boundary conditions for this system are

$$a_i^*(\boldsymbol{k}, t'') = a_i^*(\boldsymbol{k}) e^{i\omega t''}, \quad a_i(\boldsymbol{k}, t') = a_i(\boldsymbol{k}) e^{-i\omega t'} \tag{3.29}$$

$$(i=1,\ 2),$$
$$\partial_k \mathcal{A}_k(\boldsymbol{x},\ t)=0 \quad \text{for} \quad t=t';\ t=t''.$$

The boundary conditions for $\mathcal{A}_0$ follow from the system (3.27) itself and have the form

$$\partial_0 \mathcal{A}_0=0, \quad t=t',\ t=t''. \tag{3.30}$$

The solution of the system (3.27) has the form

$$\mathcal{A}_l^T(x)=\mathcal{A}_l^{T_0}(x)+\int \tilde{D}(x,\ y)\, \mathcal{J}_l^T(y)\, dy, \tag{3.31}$$

where

$$A_l^{bT_0}(x)=\sum_{i=1,\ 2} \frac{1}{(2\pi)^{3/2}} \int e^{ikx-i\omega t} a_i^b(\boldsymbol{k})\, u_l^i(\boldsymbol{k})+$$
$$+e^{-ikx+i\omega t} a_i^{b*}(\boldsymbol{k})\, u_l^i(-\boldsymbol{k}) \frac{d^3k}{\sqrt{2\omega}}, \tag{3.32}$$

and the vectors $u_l^i(\boldsymbol{k})$ are the ones introduced previously in (2.43). The Green function $\tilde{D}(x,\ y)$ has the form

$$\tilde{D}(x,\ y)=D_c(x-y)\, \theta(t''-y_0)\, \theta(y_0-t'), \tag{3.33}$$

and as $t' \to -\infty$, $t'' \to +\infty$, transforms into the causal Green function $D_c(x-y)$. The remaining components $\mathcal{A}_l^L(x)$ and $\mathcal{A}_0(x)$ are given by the formulas

$$\mathcal{A}_0(x)=\int D_2(x,\ y)\, \mathcal{J}_0(y)\, dy; \quad \mathcal{A}_l^L=\int D_2(x,\ y)\, \mathcal{J}_l^L(y)\, dy, \tag{3.34}$$

where $D_2(x,\ y)$ is the Green function of the d'Alembertian operator with the boundary conditions

$$\partial_0 \alpha\,|_{t=t''}=\partial_0 \alpha\,|_{t=t'}=0. \tag{3.35}$$

This function has the form

$$D_2(x,\ y)=$$
$$=\frac{1}{(2\pi)^3} \int e^{ik(x-y)} \frac{\cos[|\boldsymbol{k}|(x_0-t')] \cos[|\boldsymbol{k}|(y_0-t'')]}{|\boldsymbol{k}| \sin[|\boldsymbol{k}|(t''-t')]}\, d^3k. \tag{3.36}$$
$$x^0 \leqslant y^0.$$

At $x_0 \geq y_0$, $D_2(x, y)$ is defined by the symmetry condition. Combining the formulas (3.33), (3.36), and (3.26), we obtain the Green function in the Lorentz gauge for a finite time interval consistent with the Coulomb boundary conditions.

Let us now try to pass to the limit as $t'' \to \infty$, $t' \to -\infty$ in the expressions obtained. A limit for the function $\tilde{D}(x, y)$ exists, and it coincides with the causal Green function $D_c(x - y)$. This is in agreement with the fact that the three-dimensionally transverse components of $A_l^T(x)$ correspond to physical polarizations.

The functions $D_1(x, y)$ and $D_2(x, y)$ have no limits as $t'' \to +\infty$, $t' \to -\infty$. At the same time the limit of the integral (3.13), defining the S-matrix in the Lorentz gauge, must exist, since by construction this integral is equal to the Coulomb integral (2.46), for which a limit exists. This means that under expansion of the S-matrix (3.13) in the perturbation-theory series the total contribution of the functions D_1 and D_2 tends to a definite limit. Formally, the simplest way to calculate this limit is to regularize the functions D_1 and D_2 in the same manner, for example, by adding an infinitesimal imaginary part to the integration variable $\boldsymbol{k}^2$. As a result of such a regularization, the oscillating exponentials in the integrands for D_1 and D_2 will become either increasing or decreasing at large $|t'|$, $|t''|$, and the limit will exist. It is most convenient to assume that $\boldsymbol{k}^2$ has a negative imaginary part $-i0$, since in this case the limits of the functions D_1 and D_2 coincide with the causal function $D_c(x)$, and for the full Green function in the Lorentz gauge we obtain the manifestly covariant expression

$$D^L_{\mu\nu}(x) = -\frac{1}{(2\pi)^4} \int \left(g^{\mu\nu} - \frac{k_\mu k_\nu}{k^2 + i0} \right) \frac{1}{k^2 + i0} e^{-ikx}\, dk. \quad (3.37)$$

At the same time, the Green functions figuring in the expansion of the determinant $\det M_L$ in the perturbation-theory series also become causal, and the determinant itself becomes a complex-valued functional of $\mathscr{A}_\mu$.

We may once more emphasize here that the specific regularization used herein is not the only possible one. For instance, substituting $\boldsymbol{k}^2 \to \boldsymbol{k}^2 + i0$, we would obtain antichronological Green functions for non-physical polarizations, and the imaginary part of the determinant would change sign. The Green function $D^L_{\mu\nu}$, however, would then lose its manifest covariance.

The rather lengthy reasoning given has led us to the following answer to the question put above: all circuits of the poles of the Green functions may be considered to be Feynman ones, that is, we must interpret $1/k^2$ as $(k^2 + i0)^{-1}$. Thus, the S-matrix in the Lorentz gauge has the form

$$S = N^{-1} \int_{\substack{\mathcal{A}_\mu \to \mathcal{A}_{\mu \,\text{in/out}}, \\ t \to \mp\infty}} \exp\left\{ i \int \frac{1}{8} \operatorname{tr} \mathcal{F}_{\mu\nu} \mathcal{F}_{\mu\nu}\, dx \right\} \times$$

$$\times \prod_x \Delta_L(\mathcal{A})\, \delta(\partial_\mu \mathcal{A}_\mu)\, d\mathcal{A}, \tag{3.38}$$

where $\mathcal{A}_{\mu \,\text{in/out}}$ is the solution of the equations

$$\square \mathcal{A}_\mu = 0, \quad \partial_\mu \mathcal{A}_\mu = 0, \tag{3.39}$$

parametrized by the amplitudes $a_\mu(k)$ and $a^*_\mu(k)$ such that

$$a_0 = 0, \quad k_l a_l = 0; \qquad a_0^* = 0, \quad k_l a_l^* = 0, \tag{3.40}$$

the amplitude $a_l(k)$ being given in $\mathcal{A}_{\text{in}}$ (incoming wave) and $a_l^*(k)$ in $\mathcal{A}_{\text{out}}$ (outgoing wave).

An analogous derivation of the formula (3.13) for the S-matrix could be given, proceeding from the Hamilton gauge $\mathcal{A}_0 = 0$. The only difference is that, first, the integral $\int \delta(\mathcal{A}_0^\omega)\, d\omega$ on the surface $\mathcal{A}_0 = 0$ is independent of $\mathcal{A}_\mu$, so that $\Delta_H = 1$; and second, the substitution $\mathcal{A}_\mu \to \mathcal{A}_\mu^\omega$ implies that the condition

$$\partial_0 \omega(\boldsymbol{x}, t) = 0 \quad \text{for} \quad t = t' \text{ and } t = t'', \tag{3.41}$$

must be used, so that the total propagator $D^L_{\mu\nu}$ is constructed with the Green function D_2. The final result for the S-matrix obviously coincides with the previous one.

The obtained formula (3.38) is not the only possible relativistic-invariant expression for the S-matrix. Integration over gauge-equivalent classes may be performed in other ways than by selecting a representative from each class with the help of the gauge condition. Analysis of our passage from the Coulomb gauge to the Lorentz gauge reveals that in the formula (3.2) it is not necessary to use a functional of the δ-function type as an integrand. Instead, one may take any functional $B(\mathcal{A})$ which is not gauge-invariant, but for which the integral

$$\Delta_B^{-1}(\mathscr{A}) = \int [B(\mathscr{A}_\mu^\omega)] \prod_x d\omega. \tag{3.42}$$

converges. As a result, a path integral for the S-matrix appears in which $\delta(\partial_\mu \mathscr{A}_\mu)$ det $M_L(\mathscr{A})$ is replaced by $\triangle_B(\mathscr{A}_\mu)B(\mathscr{A}_\mu)$. Taking the functional

$$\exp\left\{-\frac{i}{4\alpha}\int \operatorname{tr}(\partial_\mu \mathscr{A}_\mu)^2\, dx\right\}, \tag{3.43}$$

for $B(\mathscr{A})$, we obtain a family of free Green functions $D^\alpha_{\mu\nu}$,

$$D^\alpha_{\mu\nu}(x) = -\frac{1}{(2\pi)^4}\int e^{-ikx}\left\{g_{\mu\nu} - \frac{k_\mu k_\nu(1-\alpha)}{k^2+i0}\right\}\frac{1}{k^2+i0}, \tag{3.44}$$

which contains the most widely used special cases: at $\alpha = 0$ we come back to the Lorentz gauge, and at $\alpha = 1$ we obtain the diagonal Green function.

We shall give some formal reasoning which realizes this program as simply as possible. First, let us pass from the Lorentz gauge to the generalized Lorentz gauge:

$$\partial_\mu \mathscr{A}_\mu(x) = a(x), \tag{3.45}$$

where $a(x)$ is an arbitrary matrix, using the same reasoning as when passing from the Coulomb to the Lorentz gauge. The corresponding functional $\triangle_a(\mathscr{A})$, given by the formula

$$[\triangle_a(\mathscr{A})]^{-1} = \int \prod_x \delta\left[\partial_\mu \mathscr{A}_\mu^\omega - a(x)\right] d\omega, \tag{3.46}$$

coincides on the surface

$$\partial_\mu \mathscr{A}_\mu = a(x) \tag{3.47}$$

with the functional det M, where the operator M is given by the formula (3.15). Thus the generating functional (3.38) for the S-matrix is identically rewritten as

$$S = N^{-1} \int\limits_{\mathcal{A} \to \mathcal{A}_{\substack{\text{in} \\ \text{out}}}} \exp\left\{ i \int \frac{\text{tr}}{8} \mathcal{F}_{\mu\nu} \mathcal{F}_{\mu\nu} \, dx \right\} \times$$

$$\times \prod_x \delta\left(\partial_\mu \mathcal{A}_\mu - a(x)\right) \det M \, d\mathcal{A}. \qquad (3.48)$$

Since the initial functional does not depend on a, we can integrate it over $a(x)$ with the weight

$$\exp\left\{ -i \frac{\text{tr}}{4\alpha} \int a^2(x) \, dx \right\}, \qquad (3.49)$$

which leads only to a change in the normalization constant N. Performing the integration, we obtain the generating functional for the S-matrix in the form

$$S = N^{-1} \int\limits_{\mathcal{A} \to \mathcal{A}_{\substack{\text{in} \\ \text{out}}}} \exp\left\{ i \int \text{tr} \left[\frac{1}{8} \mathcal{F}_{\mu\nu} \mathcal{F}_{\mu\nu} - \frac{1}{4\alpha} (\partial_\mu \mathcal{A}_\mu)^2 \right] dx \right\} \times$$

$$\times \prod_x \det M \, d\mathcal{A}. \qquad (3.50)$$

Extending the notion of the gauge condition, we shall call this functional the S-matrix in the α-gauge.

Expansion of this functional in the perturbation-theory series generates the diagram technique with the Green functions (3.44). In order to make this reasoning quite rigorous it is necessary, as above, to deal more carefully with the boundary conditions. We shall not do this here and shall restrict ourselves to pointing out that all the Green functions may be chosen to be causal. The equivalence of the S-matrix in various gauges will be discussed in greater detail in the next chapter, in connection with the problem of renormalization.

It is possible to introduce gauges of an even more generalized form, for which the longitudinal part of the Green function of the Yang–Mills field is an arbitrary function of k^2. For this it is sufficient to use as the functional $B(A)$ an expression of the type $\exp\{-(i/4\alpha) \, \text{tr}[\, f(\Box)\partial_\mu A_\mu]^2 \, dx\}$, where $f(\Box)$ is an arbitrary function of the d'Alembertian operator. All the reasoning given above for the case of $f \equiv 1$ is applicable without any change to this case. Gauges of this type will be used further on in the discussion of regularization and renormalization.

The expression (3.50) for the S-matrix contains the nonlocal functional $\det M$ and therefore does not look like the familiar integral of the Feynman functional $\exp\{i \times \text{action}\}$ over all fields. We may, however, use for $\det M$ the integral representation

$$\det M = \int \exp\left\{ i \int \bar{c}^a(x)\, M^{ab} c^b(x)\, dx \right\} \prod_x d\bar{c}\, dc, \tag{3.51}$$

where $\bar{c}(x)$ and $c(x)$ are anticommuting scalar functions (generators of the Grassman algebra). The boundary conditions for $\bar{c}$, c leading to the previous choice of M have the Feynman form

$$\begin{aligned} d^{a*}_{\text{out}}(\boldsymbol{k}) &= 0, \quad g^{a*}_{\text{out}}(\boldsymbol{k}) = 0, \\ d^{a}_{\text{in}}(\boldsymbol{k}) &= 0, \quad g^{a}_{\text{in}}(\boldsymbol{k}) = 0, \end{aligned} \tag{3.52}$$

where d, g, d^*, g^* are given by usual formulas

$$\begin{aligned} c^a(\boldsymbol{x}, t)_{\substack{\text{in}\\\text{out}}} &= \frac{1}{(2\pi)^{3/2}} \int \Big[e^{i\boldsymbol{kx} - i\omega t} d^a_{\substack{\text{in}\\\text{out}}}(\boldsymbol{k}) + \\ &\qquad + e^{-i\boldsymbol{kx} + i\omega t} g^{a*}_{\substack{\text{in}\\\text{out}}}(\boldsymbol{k}) \Big] \frac{d^3 k}{\sqrt{2\omega}}, \\ \bar{c}^a(\boldsymbol{x}, t)_{\substack{\text{in}\\\text{out}}} &= \frac{1}{(2\pi)^{3/2}} \int \Big[e^{i\boldsymbol{kx} - i\omega t} g^a_{\substack{\text{in}\\\text{out}}}(\boldsymbol{k}) + \\ &\qquad + e^{-i\boldsymbol{kx} + i\omega t} d^{a*}_{\substack{\text{in}\\\text{out}}}(\boldsymbol{k}) \Big] \frac{d^3 k}{\sqrt{2\omega}}. \end{aligned} \tag{3.53}$$

Using this representation, we rewrite the formula (3.50) for S as

$$\begin{aligned} S = N^{-1} \int \exp\Big\{ i \int \operatorname{tr} \Big[\frac{1}{8} \mathscr{F}_{\mu\nu} \mathscr{F}_{\mu\nu} - \frac{1}{4\alpha} (\partial_\mu \mathscr{A}_\mu)^2 - \\ - \frac{1}{2} \bar{c} (\Box c - g \partial_\mu [\mathscr{A}_\mu, c]) \Big] dx \Big\} \prod_x d\mathscr{A}\, d\bar{c}\, dc, \end{aligned} \tag{3.54}$$

where as $t \to \pm\infty$

$$\mathscr{A}_\mu \to \mathscr{A}_{\mu \substack{\text{in}\\\text{out}}}, \quad c \to c_{\substack{\text{in}\\\text{out}}}, \quad \bar{c} \to \bar{c}_{\substack{\text{in}\\\text{out}}}, \quad c = c^a t^a. \tag{3.55}$$

At the cost of introducing the fictitious fields $\bar{c}$, c we have succeeded in taking the relativity principle into accounc in such a way that the S-matrix is represented in the form of an integral of $\exp\{i \times \text{action}\}$, where the action is local and has a nondegenerate quadratic form, and integration is performed over all fields. This allows us to develop a perturbation theory for the functional (3.54), as was done in the previous chapter for the case of a scalar field, proceeding from the Gaussian integral.

With this aim we introduce the generating functional for the Green functions

$$Z(J_\mu, \bar{\xi}, \xi) = N^{-1} \int \exp\Big\{ i \int \mathcal{L}_\alpha(x) + [J_\mu^a A_\mu^a + \bar{\xi}^a c^a + \bar{c}^a \xi^a]\, dx \Big\} \prod_x d\mathcal{A}\, d\bar{c}\, dc =$$
$$= \exp\Big\{ iV\Big\{ \frac{1}{i}\frac{\delta}{\delta J_\mu}, \frac{1}{i}\frac{\delta}{\delta \bar{\xi}}, \frac{1}{i}\frac{\delta}{\delta \xi} \Big\}\Big\} \times$$
$$\times \exp\Big\{ -\frac{i}{2} \int [J_\mu^a(x)\, D_{\mu\nu}^{aab}(x - y)\, J_\nu^b(y) + 2\bar{\xi}^a(x)\, D^{ab}(x-y)\, \xi(y)]\, dx\, dy \Big\}, \quad (3.56)$$

where J_μ^a, $\bar{\xi}^a$, ξ^a are the sources of the fields A_μ^a, c^a, $\bar{c}^a$, the ξ^a and $\bar{\xi}^b$ anticommuting with each other and with the fields $\bar{c}^a$, c^b, and

$$V(\mathcal{A}_\mu, \bar{c}, c) = \frac{1}{8} \operatorname{tr} \int \{2g(\partial_\nu \mathcal{A}_\mu - \partial_\nu \mathcal{A}_\nu)[\mathcal{A}_\mu, \mathcal{A}_\nu] + g^2([\mathcal{A}_\mu, \mathcal{A}_\nu])^2 + g\bar{c}\partial_\mu[\mathcal{A}_\mu, c]\}\, dx; \quad (3.57)$$

the derivatives with repect to $\bar{\xi}$ are considered left-handed, and with respect to ξ right-handed. In the integral (3.56) all the integration variables satisfy the Feynman boundary conditions. The expansion of the functional Z in the perturbation-theory series generates the diagram technique. We list its elements, using from the beginning the momentum representation.

1. The propagator of the vector particles:

$$p \overset{\mu,a\qquad \nu,b}{\sim\!\sim\!\sim} =$$

$$= D_{\mu\nu}^{ab}(p) = -\frac{\delta^{ab}}{p^2 + i0}\Big(g_{\mu\nu} - \frac{p_\mu p_\nu}{p^2 + i0}(1 - \alpha) \Big). \quad (3.58)$$

2. The self-interaction of vector particles:

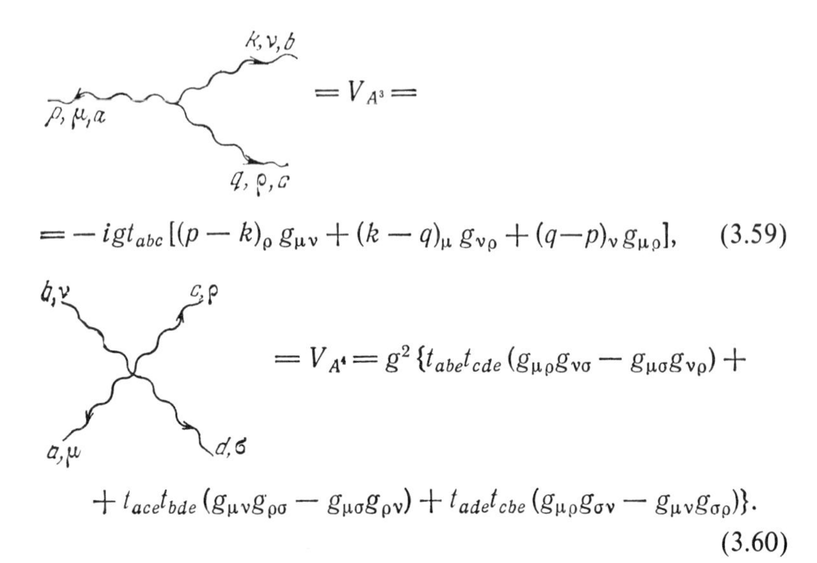

$$= V_{A^3} =$$

$$= -igt_{abc}[(p-k)_\rho g_{\mu\nu} + (k-q)_\mu g_{\nu\rho} + (q-p)_\nu g_{\mu\rho}], \qquad (3.59)$$

$$= V_{A^4} = g^2\{t_{abe}t_{cde}(g_{\mu\rho}g_{\nu\sigma} - g_{\mu\sigma}g_{\nu\rho}) +$$

$$+ t_{ace}t_{bde}(g_{\mu\nu}g_{\rho\sigma} - g_{\mu\sigma}g_{\rho\nu}) + t_{ade}t_{cbe}(g_{\mu\rho}g_{\sigma\nu} - g_{\mu\nu}g_{\sigma\rho})\}. \qquad (3.60)$$

3. The propagator of the fictitious c-particles:

$$a \text{- - - - - - - - -} b = D^{ab} = -\frac{\delta^{ab}}{p^2 + i0}. \qquad (3.61)$$

4. The interaction vertex of the fictitious c-particles with the Yang–Mills field:

$$b, k;\ a, q;\ \mu, d, p \quad = V_{\bar{c}cA} = \frac{-ig}{2} t_{abc}(k-q)_\mu. \qquad (3.62)$$

Each diagram involving these elements defines the contribution to the Green functions $G_{n,m}(k_1, \ldots, k_n | p_1, \ldots, p_m)$ with n external legs for the vector particles and m legs for the fictitious c-particles. The contribution of a given diagram enters with the factor

$$\frac{1}{r}\left(\frac{i}{(2\pi)^4}\right)^{l-V-1}(-1)^s, \qquad (3.63)$$

where V is the number of vertices, l is the number of internal lines, r is the order of the symmetry group of the diagram, and s is the number of closed loops of fictitious particles.

The S-matrix is calculated with the Green functions, using the reduction formulas:

$$
\begin{aligned}
S_{i_1 \ldots i_n, j_1 \ldots j_m}&(k'_1 \ldots k'_n, k_1 \ldots k_m) = \\
&= k_1'^2 \ldots k_n'^2 k_1^2 \ldots k_m^2 \theta(k'_{10}) \ldots \theta(k'_{n0}) \times \\
&\times \theta(-k_{10}) \ldots \theta(-k_{m0}) u^{i_1}_{\mu_1} \ldots u^{i_n}_{\mu_n} \times \qquad (3.64) \\
\times G_{\mu_1 \ldots \mu_n \nu_1 \ldots \nu_m}&(k'_1 \ldots k'_n, k_1 \ldots k_m) u^{j_1}_{\nu_1} \ldots u^{j_m}_{\nu_m} \Big|_{\substack{k_i^2 = 0 \\ k_j'^2 = 0}} .
\end{aligned}
$$

In this cumbersome formula we have multiplied each external vector line with a momentum k by k^2 and the polarization vector $u^i_\mu = (0, u^i_l)$, $i = 1, 2$, and then passed to the mass shell $k^2 = 0$, assuming that $k_{i0} > 0$ for each incoming particle and $k_{i0} < 0$ for each outgoing particle. The fictitious particles have no corresponding external lines and enter into the S-matrix only by means of closed loops.

3.4. INTERACTION WITH FIELDS OF MATTER

The consideration of matter fields $\psi(x)$ interacting with the Yang–Mills field $\mathscr{A}_\mu(x)$ does not give rise to new difficulties in the quantization problem. The action of the gauge group on the field $\mathscr{A}_\mu(x)$ is described by the same formulas as without the matter fields. Therefore, the gauge condition, imposed only on the field $\mathscr{A}_\mu(x)$, fixes the choice of the representatives in the classes of gauge-equivalent fields $\mathscr{A}_\mu(x)$, $\psi(x)$. This means that in the definition of the S-matrix for these fields one may in the corresponding path integral integrate over the fields ψ with a measure already calculated beforehand (for instance, $\Pi_x d\varphi(x)$ for a scalar field, $\Pi_x d\bar{\psi}(x)\, d\psi(x)$ for a spinor field), and as a measure for the fields $\mathscr{A}_\mu$ take one of the measures calculated in the previous section for the Yang–Mills field in vacuum. A rigorous derivation must be based on the Hamiltonian formulation of the dynamics, and only repeats the reasoning, which already has been presented more than once.

At the same time the gauge condition can be imposed also on the matter field ψ. This is convenient, specifically, for the quantization of models with spontaneous symmetry breaking. An example of such a condition will be given below.

We shall start with the example of interacting Yang–Mills and spinor fields. The Lagrangian

$$\mathcal{L} = \frac{1}{8g^2} \operatorname{tr} \{\mathcal{F}_{\mu\nu} \mathcal{F}_{\mu\nu}\} + i\bar{\psi}\gamma_\mu \nabla_\mu \psi - m\bar{\psi}\psi \tag{4.1}$$

is invariant under the gauge transformation

$$\psi(x) \to \Gamma(\omega)\psi(x); \quad \mathcal{A}_\mu \to \mathcal{A}_\mu^\omega = \omega \mathcal{A}_\mu \omega^{-1} + \partial_\mu \omega \omega^{-1}. \tag{4.2}$$

The Hamilton gauge condition

$$\mathcal{A}_0 = 0 \tag{4.3}$$

is admissible and leads to the equations of motion in the generalized Hamiltonian formulation (with a natural modification taking account of the anticommutativity of the fields $\bar{\psi}$, ψ)

$$\begin{aligned} \partial_0 \mathcal{A}_k &= \frac{\delta h}{\delta \mathcal{F}_{k0}} = \{h, \mathcal{A}\}, \quad \partial_0 \mathcal{F}_{k0} = -\frac{\delta h}{\partial \mathcal{A}_k} = \{h, \mathcal{F}_{k0}\}, \\ \partial_0 \psi &= -i\frac{\delta h}{\delta \psi^*} = \{h, \psi\}, \qquad \partial_0 \psi^* = i\frac{\delta h}{\delta \psi} = \{h, \psi^*\}, \end{aligned} \tag{4.4}$$

where $\psi^* = \bar{\psi}\gamma_0$

$$h = \int \left[i\bar{\psi}\gamma_k (\partial_k - \mathcal{A}_k)\psi + m\bar{\psi}\psi - \frac{1}{g^2} \operatorname{tr}\left(\mathcal{F}_{0k}^2 + \mathcal{F}_{ik}^2\right)\right] d^3x. \tag{4.5}$$

In addition, among the equations of motion there are the constraints

$$C^a(x) = \partial_k F_{k0}^a - t^{abc} A_k^b F_{0k}^c + i\bar{\psi}\gamma_0 \Gamma(T^a)\psi, \tag{4.6}$$

differing from (2.63) by the last term, which is constructed in terms of the matter fields. Note that this term is the 0 component of the current

$$J_\mu^a = \bar{\psi}\gamma_\mu \Gamma(T^a)\psi, \tag{4.7}$$

which is conserved in the absence of interaction. The relations

$$\begin{aligned} &\{C^a(x), C^b(y)\} = t^{abc}\delta(x-y)C^c(x), \\ &\{h, C^a\} = 0; \end{aligned}$$

$$\{C^a(x)\, A_k^b(y)\} = \delta^{ab}\partial_k\delta(x-y) - t^{acb}A_k^c\delta(x-y),$$
$$\{C^a(x),\ \psi(y)\} = \Gamma(T^a)\,\psi(x)\,\delta(x-y);$$
$$\{C^a(x),\ \bar{\psi}(y)\} = -\Gamma(T^a)\,\bar{\psi}(x)\,\delta(x-y), \tag{4.8}$$

analogous to (2.66), (2.68) and (2.69), show that $C^a(x)$ is a generator of gauge transformations, which remain after the imposition of the gauge condition $\mathscr{A}_0 = 0$. The parameters α^a of this transformation are independent of x_0.

The fields of matter enter quadratically into $C^a(x)$, so that the constraint is linearized on the solution of the free equations as $t \to \infty$

$$C^a(x) \underset{|t|\to\infty}{\sim} C_0^a(x), \tag{4.9}$$

where

$$C_0^a = \partial_k F_{k0}^a. \tag{4.10}$$

As a result, repeating the reasoning of Section 3.2, we come to the conclusion that in the quantum case, if the S-matrix is constructed with the Hamiltonian H in the large space where all the fields A_k^a, E_k^a, $\bar{\psi}^a$, ψ^a act then it commutes with the operator $C_0^a(x)$:

$$[S,\ C_0^a(x)] = 0. \tag{4.11}$$

In other words, in the presence of fields of matter only quanta of these fields and three-dimensionally transverse quanta of the Yang–Mills fields are scattered.

Note that these conclusions are based, as before, on the linearization of the constraint $C^a(x)$ at large values of time. In the framework of perturbation theory such a linearization seems quite convincing, and we assume it to take place. At the same time we cannot exclude that beyond the scope of perturbation theory linearization does not occur. The models of quark confinement are based exactly on this circumstance.

Coming back to our S-matrix, we write it in the form of a path integral

$$S = \int\limits_{\substack{\mathscr{A}\to\mathscr{A}_{\substack{\text{in}\\ \text{out}}} \\ \psi\to\psi_{\substack{\text{in}\\ \text{out}}}}} \exp\left\{i\int \mathscr{L}(x)\,dx\right\} \prod_x \delta(\mathscr{A}_0)\,d\mathscr{A}\,d\bar{\psi}\,d\psi \tag{4.12}$$

and apply to this integral the transformations already described in Section 3.3:

1. Integration over $\mathscr{F}_{k0}$.
2. Transfer to the generalized Lorentz gauge

$$\Phi_L = \partial_\mu \mathscr{A}_\mu(x) + a(x) = 0$$

using the formula

$$\delta(\mathscr{A}_0)\, d\mathscr{A}_\mu \to \Delta_L(\mathscr{A})\, \delta(\partial_\mu \mathscr{A}_\mu + a)\, d\mathscr{A}_\mu. \tag{4.13}$$

3. Integration over the auxiliary function $a(x)$ with the Gaussian weight $\exp\{-(i/4\alpha)\,\mathrm{tr}\int a^2(x)\,dx\}$.

We shall obtain expressions both for the S-matrix and for the generating functional of the Green functions in the α-gauge, which differ from the formulas (3.54), (3.56) only by the presence of fields $\bar{\psi}$, ψ, in the Lagrangian and in the terms with sources.

Besides the already introduced elements $G_{\mu\nu}$, G, V_{A^3}, V_{A^4}, $V_{c\bar{c}A}$, the diagram technique contains also a spinor line

$$p \;\text{———————}\; = S = \frac{m + p_\mu \gamma_\mu}{m^2 - p^2 - i0} \tag{4.14}$$

and vertex

$$= V_{\bar{\psi}\psi A} = g\gamma_\mu T(t_a). \tag{4.15}$$

Due to the already noted special features of integration over Fermi fields, each fermion cycle gives an additional factor (-1). In the reduction formulas the spinor legs are multiplied by $(2\pi)^{-3/2}\bar{u}_i(\gamma_\mu k_\mu - m)\theta(-k_0)$ for the incoming particle and by $(2\pi)^{-3/2}(\gamma_\mu k_\mu - m)u_i\theta(k_0)$ for the outgoing particle. For antiparticles u_i must be replaced by v_i and $\theta(k_0)$ by $\theta(-k_0)$.

The second example to be considered is the model with spontaneously broken symmetry. As the gauge group we choose the $SU(2)$ group and let φ be the scalar field in the isospinor representation

$$\varphi = \begin{pmatrix} \varphi_1 \\ \varphi_2 \end{pmatrix}, \quad \varphi^+ = (\varphi_1^*, \varphi_2^*). \tag{4.16}$$

The Lagrangian of the model (1.3.25) is rewritten in the first-order formalism as

$$\mathcal{L} = \mathcal{L}_{YM}(A_\mu, F_{\mu\nu}) + \varphi_\mu^+ \nabla_\mu \varphi + (\nabla_\mu \varphi)^+ \varphi_\mu - \varphi_\mu^+ \varphi_\mu - \\ - \lambda (\varphi^+ \varphi - \mu^2)^2, \qquad (4.17)$$

where we have introduced auxiliary vector fields φ_μ, $\overset{+}{\varphi_\mu}$, and

$$\nabla_\mu \varphi = \partial_\mu \varphi + \frac{ig}{2} A_\mu^a \tau^a \varphi, \qquad (4.18)$$

τ^a are the Pauli matrices.

As in the case of the Yang–Mills field in vacuum, the equations of motion allow us to exclude a part of the variables:

$$\mathcal{F}_{ik} = \partial_k \mathcal{A}_i - \partial_i \mathcal{A}_k + g[\mathcal{A}_i, \mathcal{A}_k]; \qquad (4.19)$$

$$\varphi_k = \nabla_k \varphi. \qquad (4.20)$$

Then the Lagrangian takes the form characteristic of the generalized Hamilton system

$$\mathcal{L} = F_{0k}^a \partial_0 A_k^a + \overset{+}{\varphi_0} \partial_0 \varphi + \partial_0 \overset{+}{\varphi} \varphi_0 - h(F_{0k}, \mathcal{A}_k, \varphi_0, \varphi) + \\ + A_0^a \left(\partial_k F_{0k}^a - g \varepsilon^{abc} A_k^b F_{0k}^c + \frac{ig}{2} (\varphi_0^+ \tau^a \varphi - \varphi^+ \tau^a \varphi_0) \right), \qquad (4.21)$$

where h is the Hamiltonian, the explicit form of which we do not write out.

As is seen, the pairs F_{0k}^a, A_k^a and φ_0, φ play the role of conjugate canonical variables, A_0^a is the Lagrangian multiplier, and

$$C^a = -\partial_k F_{0k}^a + g \varepsilon^{abc} A_k^b F_{0k}^c + \frac{ig}{2} (\varphi_0^+ \tau^a \varphi - \varphi^+ \tau^a \varphi_0) \qquad (4.22)$$

is the constraint. It is easy to verify that commutation conditions such as (2.10) are fulfilled, so we can use the general formulas from Section 3.2 for quantization of this model.

As was pointed out in Chapter One, the fields

$$A_\mu^a = 0; \quad \varphi = \begin{pmatrix} 0 \\ \mu \end{pmatrix}. \qquad (4.23)$$

correspond to a stable vacuum configuration. Therefore, before passing to the quantization it is convenient to make a shift

$$\varphi \to \varphi - \binom{0}{\mu}. \tag{4.24}$$

As a result of such a shift, the constraint (4.22) takes the form

$$C^a = -\partial_k F^a_{0k} + g\varepsilon^{abc} A^b_k F^c_{0k} + m_1 B^a_0 + \\ + \frac{1}{2} g \left(\sigma B^a_0 - B^a \sigma_0 - \varepsilon^{abc} B^b B^c_0\right), \tag{4.25}$$

where the following notation is introduced:

$$\varphi^1 = \frac{iB^1 + B^2}{\sqrt{2}}; \quad \varphi^2 = \mu + \frac{1}{\sqrt{2}} (\sigma - iB^3), \quad m_1 = \frac{\mu g}{\sqrt{2}}, \tag{4.26}$$

and B^a_0 and σ_0 stand for the corresponding canonical momenta.

We see that the constraint contains the term B^a_0 linearly; therefore it is natural to choose as an additional condition (it is also the gauge condition)

$$B^a = 0. \tag{4.27}$$

Indeed, the matrix of the Poisson brackets

$$\{C^a(x),\ B^b(y)\} = \left(m_1 + \frac{g}{2} \sigma(x)\right) \delta^{ab} \delta(x - y) + \ \ldots, \tag{4.28}$$

where $\cdots$ indicates terms vanishing at $B^a = 0$, is nondegenerate in the framework of perturbation theory, that is, at $|g\sigma| \ll m_1$.

In the gauge $B^a = 0$ the quadratic form of the Hamiltonian h and the linear form of the constraint C take the form

$$h_0 = \frac{1}{2} (F^a_{0k})^2 + \frac{1}{4} (\partial_i A^a_k - \partial_k A^a_i)^2 + \frac{m_1^2}{2} (A^a_k)^2 + \frac{1}{2} (B^a_0)^2 + \\ + \frac{1}{2} \sigma_0^2 + \frac{1}{2} (\partial_k \sigma)^2 + \frac{m_2^2}{2} \sigma^2, \quad m_2 = 2\lambda\mu, \tag{4.29}$$

$$C^a_0 = -\partial_k F^a_{0k} + m_1 B^a_0. \tag{4.30}$$

As a free Hamiltonian h^*_0 defining the spectrum of particles in the framework of perturbation theory, we must choose an expression which is obtained by substituting into h the solution of the equation of constraint $C_0 = 0$:

$$h_0^* = \frac{1}{2}(F_{0k}^a)^2 + \frac{1}{2m_1^2}(\partial_k F_{0k}^a)^2 + \frac{1}{4}(\partial_i A_k^a - \partial_k A_i^a)^2 + \\ + \frac{m_1^2}{2}(A_k^a)^2 + \frac{1}{2}\sigma_0^2 + \frac{1}{2}(\partial_k \sigma)^2 + \frac{m_2^2}{2}\sigma^2. \quad (4.31)$$

The expression $\int h_0^* \, d^3x$ is diagonalized by the substitution

$$A_l^b(\boldsymbol{x}) = (2\pi)^{-\frac{3}{2}} \sum_{i=1}^{3} \int \Big(e^{i\boldsymbol{k}\boldsymbol{x}} a_i^b(\boldsymbol{k})\, e_l^i(\boldsymbol{k}) + \\ + e^{-i\boldsymbol{k}\boldsymbol{x}} a_i^{*b}(\boldsymbol{k})\, e_l^i(-\boldsymbol{k})\Big) \frac{d^3k}{\sqrt{2\omega_1}}; \quad (4.32)$$

$$F_{0l}^b(\boldsymbol{x}) = (2\pi)^{-\frac{3}{2}} \sum_{i=1}^{3} \int \Big(e^{i\boldsymbol{k}\boldsymbol{x}} a_i^b(\boldsymbol{k})\, \tilde{e}_l^i(\boldsymbol{k}) - \\ - e^{-i\boldsymbol{k}\boldsymbol{x}} a_i^{*b}(\boldsymbol{k})\, \tilde{e}_l^i(-\boldsymbol{k})\Big) \frac{1}{i} \sqrt{\frac{\omega_1}{2}}\, d^3k, \quad (4.33)$$

(where $e_l^1 = \tilde{e}_l^1$ and $e_l^2 = \tilde{e}_l^2$ are two arbitrary orthonormalized vectors, orthogonal to the vector $\boldsymbol{k}$, and

$$e_l^3 = \frac{k_l}{|\boldsymbol{k}|} \frac{\omega_1}{m_1}; \quad \tilde{e}_l^3 = \frac{k_l}{|\boldsymbol{k}|} \frac{m_1}{\omega_1}, \quad \omega_1 = \sqrt{\boldsymbol{k}^2 + m_1^2}, \quad (4.34)$$

for the vector field) and by the standard substitution for the scalar field σ. As a result, the free Hamiltonian takes the form

$$\int h_0^* \, d^3x = \int d^3k \left(\sum_{i=1}^{3} a_i^{b*} a_i^b \omega_1 + a_\sigma^* a_\sigma \omega_2\right), \quad \omega_2 = \sqrt{\boldsymbol{k}^2 + m_2^2}. \quad (4.35)$$

As is seen from the given computations, the spectrum consists of three massive vector particles and one massive scalar particle.

We can now pass to the discussion of the S-matrix. In the generalized Hamiltonian formulation the expression for the kernel of the S-matrix in terms of the path integral can be written as

$$S = \lim_{\substack{t'' \to \infty \\ t' \to -\infty}} \int \exp\left\{ \frac{i}{2} \int d^3k \left(\sum_{i=1}^{3} (a_i^{b*}(\boldsymbol{k}, t'')\, a_i^b(\boldsymbol{k}, t'') + \right.\right.$$
$$+ a_i^{b*}(\boldsymbol{k}, t')\, a_i^b(\boldsymbol{k}, t')) + a_\sigma^*(\boldsymbol{k}, t'')\, a_\sigma(\boldsymbol{k}, t'') +$$
$$+ a_\sigma^*(\boldsymbol{k}, t')\, a_\sigma(\boldsymbol{k}, t')) +$$
$$+ i \int d^3x \int_{t'}^{t''} dt \left(\frac{1}{2} (F_{0k}^a \dot{A}_k^a - \dot{F}_{0k}^a A_k^a + \sigma_0 \dot{\sigma} - \dot{\sigma}_0 \sigma) - \right.$$
$$\left.\left. - h(F_{0k}, A_k, B_0, B, \sigma_0, \sigma) + A_0^a C^a \right) \right\} \prod_x \delta(B^a) \left(m_1 + \frac{g\sigma}{2} \right)^3 \times$$
$$\times \prod_x dF_{0k}\, dA_k\, d\sigma_0\, d\sigma\, dB_0\, dB\, dA_0, \qquad (4.36)$$

$$a_i^{*b}(\boldsymbol{k}, t'') = a_i^{*b}(\boldsymbol{k})\, e^{i\omega_1 t''}, \quad a_i^b(\boldsymbol{k}, t') = a_i^b(\boldsymbol{k})\, e^{-i\omega_1} \qquad (4.37)$$

$$a_\sigma^*(\boldsymbol{k}, t'') = a_\sigma^*(\boldsymbol{k})\, e^{i\omega_2 t''}; \qquad a_\sigma(\boldsymbol{k}, t') = a_\sigma(\boldsymbol{k})\, e^{-i\omega_2 t'}. \qquad (4.38)$$

Here the variables $a_i^b(\mathbf{k}, t)$, $a_i^{b*}(\mathbf{k}, t)$ are related to $A_k^b(\mathbf{x}, t)$, $F_{0k}^b(\boldsymbol{x}, t)$ by formulas like (4.32), (4.33):

$$A_l^b(\boldsymbol{x}, t) =$$
$$= \left(\frac{1}{2\pi} \right)^{3/2} \int \sum_{i=1}^{3} [a_i^b(\boldsymbol{k}, t)\, e^{i\boldsymbol{k}\boldsymbol{x}} e_l^i(\boldsymbol{k}) + a_i^{b*}(\boldsymbol{k}, t)\, e^{-i\boldsymbol{k}\boldsymbol{x}} e_l^i(-\boldsymbol{k})] \times$$
$$\times \frac{d^3k}{\sqrt{2\omega_1}}, \qquad (4.39)$$

$$F_{0l}^b(\boldsymbol{x}, t) =$$
$$= \left(\frac{1}{2\pi} \right)^{3/2} \int \sum_{i=1}^{3} [a_i^b(\boldsymbol{k}, t)\, e^{i\boldsymbol{k}\boldsymbol{x}} \tilde{e}_l^i(\boldsymbol{k}) - a_i^{b*}(\boldsymbol{k}, t)\, e^{-i\boldsymbol{k}\boldsymbol{x}} \tilde{e}_l^i(-\boldsymbol{k})] \times$$
$$\times \frac{1}{i} \sqrt{\frac{\omega_1}{2}}\, d^3k. \qquad (4.40)$$

Let us transform the expression (4.36) to a manifestly relativistic-invariant form, as we have done for the scalar field in Section 2.3. For this we must integrate over the variables B, σ_0, B_0, and F_{0k} and pass to the limit as $t'' \to \infty$, $t' \to -\infty$. The integration over σ_0 and the relevant transformations differ in no way from the already considered case of a scalar field, and so we shall not repeat them. Integration over the variable B_0 is performed after the shift

$$B_0^a \to B_0^a - m_1 A_0^a, \tag{4.41}$$

which separates B_0 from the other variables. The functions $B_0(x)$ and $A_0(x)$ do not take part in the boundary conditions, so that integration over B_0 leads only to a change in the normalization factor. As a result of the shift (4.41), a mass term of the form $(m_1^2/2)A_0^2$ for the field A_0 is added to the action. Integration over B_k removes the δ-function from (4.36).

In order to integrate over F_{0k}, it is necessary to make a shift

$$F_{0k}^a = F_{0k}^{a\,(1)} + \partial_0 A_k^a - \partial_k A_0^a. \tag{4.42}$$

Integrating by parts, we verify that in the new variables the quadratic form of the action, obtained from (4.36) under the already carried out transformations, takes the form

$$-\frac{1}{2}\int A_k^a F_{0k}^a\, d^3x \Big|_{t'}^{t''} + \int_{t'}^{t''} dt \int d^3x \Big[\frac{1}{4}(\partial_\mu A_\nu^a - \partial_\nu A_\mu^a)^2 +$$
$$+\frac{m_1^2}{2}(A_\mu^a)^2 - \frac{1}{2}(F_{0k}^{a\,(1)})^2\Big]. \tag{4.43}$$

Now let us perform a shift of the variable $\mathscr{A}_\mu$:

$$A_\mu^a = A_\mu^{a\,(0)} + A_\mu^{a\,(1)}, \tag{4.44}$$

where $\mathscr{A}_\mu^{(0)}$ is the solution of the free equation of motion

$$(\Box_{\mu\nu} + m_1^2 g_{\mu\nu})\mathscr{A}_\nu^{(0)} = 0; \quad \Box_{\mu\nu} = \Box g_{\mu\nu} - \partial_\mu\partial_\nu, \tag{4.45}$$

generated by the quadratic form (4.43). It may be written as

$$A_\mu^{b\,(0)} =$$
$$= \left(\frac{1}{2\pi}\right)^{3/2} \int [a_i^b(\boldsymbol{k})\, e^{-ikx} u_\mu^i(\boldsymbol{k}) + a_i^{b*}(\boldsymbol{k})\, e^{ikx} u_\mu^i(-\boldsymbol{k})]\big|_{k_0=\omega_1} \times$$
$$\times \frac{d^3k}{\sqrt{2\omega_1}}, \tag{4.46}$$

where

$$u^i_\mu = (0,\ e^i_k), \qquad i=1,\ 2; \qquad u^3_\mu = \left(i\,\frac{|\boldsymbol{k}|}{m},\ e^3_k\right), \quad (4.47)$$

and the vectors e^i_k, $i = 1, 2$, are the ones introduced in (4.32). Note that $\mathscr{A}^{(0)}_\mu$ satisfies the relation

$$\partial_\mu \mathscr{A}^{(0)}_\mu = 0, \qquad (4.48)$$

which must be satisfied by all solutions of the equation (4.45).

By integrating by parts it is possible to completely exclude $\mathscr{A}^{(0)}_\mu$ from the quadratic form of the action (4.43), as a result of which it takes the form

$$\int d^3x \left[\left(\partial_0 A^{a\,(0)}_k - \partial_k A^{a\,(0)}_0\right)\left(A^a_k - \frac{1}{2} A^{a\,(0)}_k\right) - \frac{1}{2} F^a_{0k} A^a_k \right]\Bigg|_{t'}^{t''} +$$
$$+ \int_{t'}^{t''} dt \int d^3x \left[-\frac{1}{2}\left(F^{a\,(1)}_{0k}\right)^2 + \frac{1}{4}\left(\partial_\mu A^{a\,(1)}_\nu - \partial_\nu A^{a\,(1)}_\mu\right)^2 + \right.$$
$$\left. + \frac{m_1^2}{2}\left(A^{a\,(1)}_\mu\right)^2 \right]. \qquad (4.49)$$

The terms outside of the integral may be expressed through the variables $a^b_i(k)$, $a^{b*}_i(k)$, as a result of which the first term in (4.49) will be written down in a form analogous to (2.3.40):

$$\frac{1}{i}\int d^3k \left\{ a^{b*}_i(\boldsymbol{k})\, a^b_i(\boldsymbol{k}) - \frac{1}{2}\left[a^{b*}_i(\boldsymbol{k},\ t'')\, a^b_i(\boldsymbol{k},\ t'') + \right.\right.$$
$$\left. + a^{b*}_i(\boldsymbol{k},\ t')\, a^b_i(\boldsymbol{k},\ t')\right] - \left(a^{b*}_i(\boldsymbol{k},\ t') - a^{b*}_i(\boldsymbol{k})\, e^{i\omega_1 t'}\right)^2 -$$
$$\left. - \left(a^b_i(\boldsymbol{k},\ t'') - a^b_i(\boldsymbol{k})\, e^{-i\omega_1 t''}\right)^2 \right\}. \qquad (4.50)$$

The second term in (4.49) remains finite in the limit $t' \to -\infty$, $t'' \to \infty$, provided that

$$A^b_0(\boldsymbol{x},\ t) =$$
$$= \left(\frac{1}{2\pi}\right)^{3/2} \int \left(a^b_3(\boldsymbol{k},\ t)\, e^{ikx} + a^{b*}_3(\boldsymbol{k},\ t)\, e^{-ikx}\right) i\,\frac{|\boldsymbol{k}|}{m_1}\,\frac{d^3k}{\sqrt{2\omega_1}} +$$
$$+ A^{b\,(1)}_0(\boldsymbol{x},\ t), \qquad (4.51)$$

where $A_0^{b(1)}$ is a rapidly decreasing function, as also is $F_{0k}^{b(1)}$. The relation (4.42) is compatible with this decrease and the boundary condition (4.37), (4.38) if

$$a_i^{b*}(\boldsymbol{k}, t') = e^{i\omega_1 t} a_{i,\,\mathrm{in}}^{b*}(\boldsymbol{k}) + a_{i,\,1}^{b*}(\boldsymbol{k}, t'), \qquad t' \to -\infty,$$
$$a_i^b(\boldsymbol{k}, t'') = e^{-i\omega_1 t''} a_{i,\,\mathrm{out}}^b(\boldsymbol{k}) + a_{i,\,1}^b(\boldsymbol{k}, t''), \quad t'' \to \infty, \tag{4.52}$$

where $a_{i,1}^b{}^*(\boldsymbol{k}, t')$ decreases rapidly as $t' \to -\infty$, and $a_{i,1}^b(\boldsymbol{k}, t'')$ as $t'' \to \infty$. As a result, we see that in the expression (4.36) it is possible to pass to the limit as $t'' \to \infty$, $t' \to -\infty$ if the integration variables behave asymptotically as the solutions of the free equation (4.45):

$$A_\mu^a(\boldsymbol{x}, t) = A_{\mu,\,\substack{\mathrm{in}\\ \mathrm{out}}}^a(\boldsymbol{x}, t) + A_{\mu,\,\substack{\mathrm{in}\\ \mathrm{out}}}^{a\,(1)}(\boldsymbol{x}, t), \tag{4.53}$$

where $A_{\mu,\,\mathrm{in}}^{a\,(1)}$, $A_{\mu,\,\mathrm{out}}^{a\,(1)}$ decrease rapidly as $t \to -\infty$ and $t \to \infty$ respectively, and $A_{\mu,\,\mathrm{in}}^a$ and $A_{\mu,\,\mathrm{out}}^a$ are represented in the form

$$A_{\mu,\,\substack{\mathrm{in}\\ \mathrm{out}}}^b = \left(\frac{1}{2\pi}\right)^{3/2} \int \Big[a_{i,\,\substack{\mathrm{in}\\ \mathrm{out}}}^b(\boldsymbol{k})\, e^{-ikx} u_\mu(\boldsymbol{k}) +$$
$$+ a_{i,\,\substack{\mathrm{in}\\ \mathrm{out}}}^{b*}(\boldsymbol{k})\, e^{ikx} u_\mu(-\boldsymbol{k}) \Big]\Big|_{k_0=\omega_1} \frac{d^3k}{\sqrt{2\omega_1}}, \tag{4.54}$$

where

$$a_{i,\,\mathrm{in}}^b(\boldsymbol{k}) = a_i^b(\boldsymbol{k}); \quad a_{i,\,\mathrm{out}}^{b*}(\boldsymbol{k}) = a_i^{b*}(\boldsymbol{k}). \tag{4.55}$$

No conditions are imposed on the functions $a_{i,\,\mathrm{out}}^b(\boldsymbol{k})$ and $a_{i,\,\mathrm{in}}^b{}^*(\boldsymbol{k})$.

From (4.54), (4.55), and (4.46) it follows that $A_\mu^{a(1)} = A_\mu^a - A_\mu^{a(0)}$ satisfies the Feynman boundary conditions, that is $A_\mu^{a(1)}$ does not contain incoming waves as $t \to -\infty$ and outgoing waves as $t \to \infty$. On such functions the quadratic form in (4.49) is uniquely defined as the quadratic form of the operator

$$\Box_{\mu\nu} + m_1^2 g_{\mu\nu}, \tag{4.56}$$

occurring in the equation (4.45).

Further transformations of the expression (4.36) precisely follow the reasoning of Section 2.3. As a result, we obtain for the normal symbol of the S-matrix the final manifestly invariant expression

$$S\left(A_\mu^{(0)}, \sigma^{(0)}\right) = \\ = \int\limits_{\substack{A \to A_{\substack{\text{in}\\\text{out}}} \\ \sigma \to \sigma_{\substack{\text{in}\\\text{out}}}}} \exp\left\{ i \int \mathscr{L}(x)\, dx \right\} \prod_x \left(m_1 + \frac{g}{2}\sigma \right)^3 dA_\mu\, d\sigma, \quad (4.57)$$

where

$$\mathscr{L} = -\frac{1}{4}\left(\partial_\nu A_\mu^a - \partial_\mu A_\nu^a + g\varepsilon^{abc} A_\mu^b A_\nu^c\right)^2 + \frac{m_1^2}{2} A_\mu^2 + \\ + \frac{1}{2}\partial_\mu \sigma \partial_\mu \sigma - \frac{m_2^2}{2}\sigma^2 + \frac{m_1 g}{2}\sigma A_\mu^2 + \frac{g^2}{8}\sigma^2 A_\mu^2 - \\ - \frac{g m_2}{4 m_1}\sigma^3 - \frac{g^2 m_2^2}{32 m_1^2}\sigma^4. \quad (4.58)$$

The asymptotic conditions for the fields σ have a form analogous to (2.3.47):

$$\sigma_{\substack{\text{in}\\\text{out}}}(x) = \left(\frac{1}{2\pi}\right)^{3/2} \int \left(a_{\sigma, \substack{\text{in}\\\text{out}}}(\boldsymbol{k})\, e^{-ikx} + a^*_{\sigma, \substack{\text{in}\\\text{out}}}(\boldsymbol{k})\, e^{ikx} \right)\Big|_{k_0 = \omega_2} \frac{d^3k}{\sqrt{2\omega_2}}, \quad (4.59)$$

where $a^*_{\sigma,\,\text{out}}(\boldsymbol{k}) = a^*_\sigma(\boldsymbol{k})$; $a_{\sigma,\,\text{in}}(\boldsymbol{k}) = a_\sigma(\boldsymbol{k})$. The asymptotic conditions for the fields A_μ^b are given by (4.54), (4.55).

The quadratic form in the action $\int \mathscr{L}(x)\, dx$ is defined as

$$\frac{1}{2}\int \left(A_\mu - A_\mu^{(0)}\right)\left(\square_{\mu\nu} + g_{\mu\nu} m_1^2\right)\left(A_\mu - A_\mu^{(0)}\right), \quad (4.60)$$

where the operator $(\square_{\mu\nu} + m_1^2 g_{\mu\nu})$ is supplied with Feynman boundary conditions. The generating functional for the Green functions,

$$Z(J, \eta) = \int \exp\left\{ i \int \left(\mathscr{L}(x) + J_\mu^a A_\mu^a + \sigma\eta \right) dx \right\} \times \\ \times \prod_x \left(m_1 + \frac{g}{2}\sigma \right)^3 d\mathscr{A}_\mu\, d\sigma \quad (4.61)$$

and the perturbation-theory diagram technique which follows from it, contain some new features. First, the propagator of the vector field

(4.51), which can be rewritten as

$$D^c_{\mu\nu}(x-y)=\left(\frac{1}{2\pi}\right)^4\int e^{ik(x-y)}\left(g_{\mu\nu}-\frac{k_\mu k_\nu}{m_1^2}\right)\times$$
$$\times\frac{1}{k^2-m_1^2+i0}\,d^4k, \quad (4.62)$$

has a higher degree of singularity at $x \sim y$ than the Green functions we have encountered until now. Indeed, its longitudinal part

$$\frac{k_\mu k_\nu}{m_1^2}\,\frac{1}{k^2-m_1^2+i0} \quad (4.63)$$

does not decrease at large k, so that its contribution to the propagator has a singularity of the power $\delta^{(4)}(x)$. Second, the integration measure contains the local factor det M_Φ. It may be formally written as

$$\det M_\Phi=\prod_x\left(m_1+\frac{g}{2}\sigma(x)\right)^3=$$
$$=\text{const}\cdot\exp\left\{V\int\ln\left(m_1+\frac{g\sigma(x)}{2}\right)^3\right\}dx=$$
$$=\text{const}\cdot\exp\left\{\delta^{(4)}(0)\left[-\sum_{n=1}^{\infty}\frac{(-1)^n}{n}\int\left(\frac{g\sigma(x)}{2m_1}\right)^n dx\right]\right\}, \quad (4.64)$$
$$V=\int dx=\delta^{(4)}(0).$$

In the framework of perturbation theory such an addition to the action generates new diagrams, the contribution of which is proportional to powers of $\delta^{(4)}(0)$ (of course, this expression is to be understood in the sense of a certain volume regularization). The role of these diagrams is to compensate the singular parts of other diagrams arising in the perturbation theory. Such singularities arise in the multiplication of δ-type contributions of vector particles to the Green functions.

Both indicated features of the diagram technique for the Lagrangian (4.58) shows that it contains inconvenient singularities. Therefore it is more convenient to investigate the model under consideration in the Lorentz gauge or in the α-gauge, which can be introduced in a simple manner, using already familiar methods. The role of the above-considered gauge (often called unitary) is that it gives the spectrum of particles and the asymptotic states of the model in a manifestly relativistic-invariant manner. In this sense it gives us a substitute for the Coulomb gauge of the Yang–Mills theory in vacuum.

We shall not again describe the procedure of passing to the α-gauge, since it does not require any new ideas in this case, because the field $\partial_\mu A_\mu$ does not propagate. The normal symbol of the S-matrix is given by the path integral

$$S = N^{-1} \int_{\substack{\mathcal{A} \to \mathcal{A}_{\text{in, out}} \\ \sigma \to \sigma_{\text{in, out}}}} \exp\left\{ i \int \left(\mathcal{L}(x) + \frac{1}{2\alpha} (\partial_\mu A_\mu)^2 \right) dx \right\} \times \times \det M_\alpha \prod_x d_\mu A \, d\mathcal{B} \, d\sigma, \tag{4.65}$$

where

$$\mathcal{L}(x) = -\frac{1}{4} F^a_{\mu\nu} F^a_{\mu\nu} + \frac{m_1^2}{2} A_\mu^2 + m_1 A^a_\mu \partial_\mu B^a + \frac{1}{2} \partial_\mu B^a \partial_\mu B^a + + \frac{1}{2} \partial_\mu \sigma \partial_\mu \sigma - \frac{m_2^2}{2} \sigma^2 + \frac{g}{2} A^a_\mu (\sigma \partial_\mu B^a - B^a \partial_\mu \sigma - \varepsilon^{abc} B^b \partial_\mu B^c) + + \frac{m_1 g}{2} \sigma A_\mu^2 + \frac{g^2}{8} (\sigma^2 + B^2) A_\mu^2 - \frac{g m_2^2}{4 m_1} \sigma (\sigma^2 + B^2) - - \frac{g^2 m_2^2}{32 m_1^2} (\sigma^2 + B^2)^2, \tag{4.66}$$

and

$$M_\alpha u = M u = \Box u - g \partial_\mu [\mathcal{A}_\mu, u]. \tag{4.67}$$

The asymptotic conditions for A^a_μ and σ are the same as before. The fields $\partial_\mu A^a_\mu$, B^a and the fictitious particles $\bar{c}^a$, c^a, taking part in the definition of det M, do not propagate. In the construction of the diagram technique it is convenient (but not necessary) to use the Feynman boundary conditions for them. The generating functional for the Green functions is constructed in a standard way by means of the expression (4.65) for the S-matrix.

The diagram technique in the generalized α-gauge is somewhat cumbersome owing to the presence of mixed propagators. In calculations it is more convenient to use the Lorentz gauge $\alpha = 0$. In this gauge the diagram technique has the following elements:

1. A propagator corresponds to the line of a vector particle

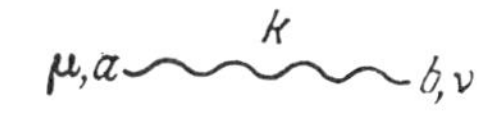

$$\mu, a \sim\sim\sim\overset{k}{\sim}\sim\sim\sim b, \nu \quad -\delta^{ab}\left[\frac{g_{\mu\nu} - k_\mu k_\nu k^{-2}}{k^2 - m_1^2 + i0}\right]. \tag{4.68}$$

2. The propagators of the fictitious particles $\bar{c}$, c and their interaction vertices with a vector particle are the same as for the Yang–Mills theory in vacuum.

3. To lines of scalar particles B^a and σ there correspond the propagators

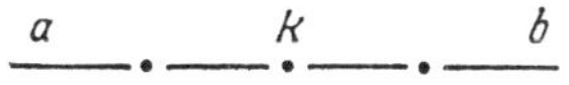

$$\overset{a}{\text{---}}\cdot\text{---}\overset{k}{\cdot}\text{---}\cdot\overset{b}{\text{---}} = \frac{\delta^{ab}}{k^2 + i0} \tag{4.69}$$

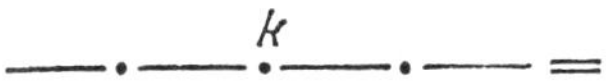

$$\text{---}\cdot\text{---}\overset{k}{\cdot}\text{---}\cdot\text{---} = \frac{1}{k^2 - m_2^2 + i0}. \tag{4.70}$$

4. There exist numerous interaction vertices of the fields $\mathscr{A}_\mu$, $\mathscr{B}$ and σ, which are readily written out according to the Lagrangian (4.66).

The reduction formulas have the usual form and we shall describe them in words. The external legs in the Green functions must be taken only for vector particles and for particles of the field σ. Each leg, corresponding to a vector field, is multiplied by $(k_j^2 - m_1^2)$ and by the polarization vector $u_\mu^i(\boldsymbol{k}_j)$. Each leg, corresponding to the field σ is multiplied by $(p_l^2 - m_2^2)$. Then it is necessary to pass on to the mass shell $k_j^2 = m_1^2, p_l^2 = m_2^2$, assuming k_0 and p_0 to be positive for outgoing particles and negative for incoming particles.

Another version of the diagram technique can be obtained if one chooses as the functional B in the formula (3.42), which removes degeneracy, the expression

$$\exp\left\{\frac{i}{2\alpha}\int(\partial_\mu A_\mu^a - \alpha m_1 B^a)^2\,dx\right\}. \tag{4.71}$$

Computations entirely analogous to those which led us to the expression for the S-matrix in the α-gauge lead to the following result:

$$S = N^{-1}\int\limits_{\substack{A\to A_{\text{in/out}}\\ \sigma\to\sigma_{\text{in/out}}}} \exp\left\{i\int\left[\mathscr{L}(x) + \frac{1}{2\alpha}(\partial_\mu A_\mu^a - \alpha m_1 B^a)^2\right]dx\right\}\times$$

$$\times\prod_x \det M_X\, dA_\mu\, dB\, d\sigma, \tag{4.72}$$

where the Lagrangian $\mathscr{L}$ is given, as before, by the formula (4.66), and the operator M_X looks as follows:

$$M_X u = (\Box + \alpha m_1^2)\, u - g\partial_\mu [\mathscr{A}_\mu, u] + \frac{\alpha g m_1}{2} [\mathscr{B}, u] + \\ + \frac{\alpha m_1 g}{2} \sigma u. \qquad (4.73)$$

As before, det M_X can be represented by an integral over the fields of fictitious particles:

$$\det M_X = \int \exp\left\{ i \int \bar{c}^a(x)\, M_X^{ab} c^b(x)\, dx \right\} \prod_x d\bar{c}\, dc. \qquad (4.74)$$

The term $m_1 A^a_\mu \partial_\mu B^a$ that is nondiagonal in the fields A_μ, B in the Lagrangian (4.66) cancels out with an analogous term in the expression fixing the gauge. As a result, mixed propagators $\overline{A^a_\mu B^b}$ are absent. The Feynman rules differ from the ones formulated above in the following points:

1. The propagator corresponding to a vector line is

$$-\delta^{ab}\left[\frac{g_{\mu\nu} - k_\mu k_\nu k^{-2}}{k^2 - m_1^2 + i0} + \frac{k_\mu k_\nu k^{-2}}{(k^2\alpha^{-1} - m_1^2 + i0)}\right]. \qquad (4.75)$$

2. The propagator corresponding to the scalar B-line is

$$\frac{\delta^{ab}}{p^2 - m_1^2\alpha + i0}. \qquad (4.76)$$

3. The propagator corresponding to fictitious particles is

$$-\frac{\delta^{ab}}{p^2 - m_1^2\alpha + i0}. \qquad (4.77)$$

4. There appear additional interaction vertices of fictitious particles with the fields B^a and σ. Their explicit forms are easy to derive from the formulas (4.73) and (4.74).

At $\alpha = 0$ these rules obviously coincide with the rules formulated above for the Lorentz gauge.

At this point we conclude the description of examples of interaction of the Yang–Mills field with fields of matter. We hope that the examples have been sufficiently typical, and that the reader will be able without difficulty to construct a diagram technique for any arbitrary model either with or without symmetry breaking.

CHAPTER 4

RENORMALIZATION OF GAUGE THEORIES

4.1. EXAMPLES OF THE SIMPLEST DIAGRAMS

The diagram technique developed in the previous chapter allows one to calculate Green functions and probabilities of scattering processes to the accuracy of any order in g. However, direct application of the rules formulated above to the calculation of diagrams containing closed loops leads to a meaningless result—the corresponding integrals diverge at large momenta. Attaching meaning to these expressions is the essence of the renormalization procedure to be studied in the present chapter.

As the simplest example, we shall consider the second-order correction to the Green function of a fictitious particle in the Yang–Mills theory with the gauge group $SU(2)$. This correction is described by the diagram in Figure 1. The corresponding analytical expression has the form

$$-\frac{g^2}{k^2+i0}\,\Sigma(k^2)\,\frac{1}{k^2+i0}, \tag{1.1}$$

where in the diagonal α-gauge ($\alpha = 1$)

$$\Sigma(k^2) = -\frac{i2k_\nu g_{\mu\nu}}{(2\pi)^4}\int \frac{dp\,(k-p)_\mu}{(p^2+i0)\,[(k-p)^2+i0]}. \tag{1.2}$$

As $p \to \infty$ this integral diverges linearly. In order to attach meaning to the integral (1.2), we first introduce an intermediate regularization,

L. D. Faddeev and A. A. Slavnov. Gauge Fields: Introduction to Quantum Theory, ISBN 0-8053-9016-2.

Copyright © 1980 by The Benjamin/Cummings Publishing Company, Advanced Book Program. All rights reserved. No part of this publication may be reproduced, stored in a retrieval system, or transmitted, in any form or by any means, electronic, mechanical photocopying, recording, or otherwise, without the prior permission of the publisher.

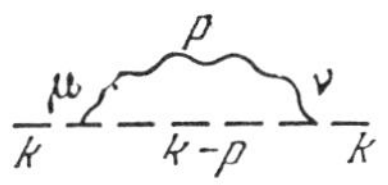

Fig. 1. Second-order correction to the Green function of a fictitious particle. The dashed line indicates the propagator of the fictitious particle and the wavy line of the Yang–Mills field.

replacing the function $(p^2 + i0)^{-1}$ by the regularized expression

$$\frac{1}{(p^2+i0)} \to \frac{1}{(p^2+i0)} - \frac{1}{p^2-\Lambda^2+i0} =$$
$$= -\int_0^{\Lambda^2} \frac{d\lambda}{(p^2-\lambda+i0)^2}. \quad (1.3)$$

As $\Lambda \to \infty$ the regularized Green function tends to the initial expression $(p^2 + i0)^{-1}$. At finite Λ the integral

$$\Sigma_\Lambda(k^2) = \frac{2ik_\nu}{(2\pi)^4} \int_0^{\Lambda^2} d\lambda \int \frac{dp\,(k-p)_\nu}{[p^2-\lambda+i0]^2\,[(k-p)^2+i0]} \quad (1.4)$$

converges. To calculate it, we use the Feynman formula

$$\frac{1}{a^2 b} = \int_0^1 \frac{2z\,dz}{[az+b(1-z)]^3}. \quad (1.5)$$

This formula allows us to combine both factors in the denominator of the integral (1.4) into one:

$$\Sigma_\Lambda(k^2) = \frac{2ik_\nu}{(2\pi)^4} \int_0^{\Lambda^2} d\lambda \int_0^1 dz 2z \times$$
$$\times \int \frac{dp\,(k-p)_\nu}{[(p^2-\lambda+i0)\,z + (k^2-2pk+p^2+i0)\,(1-z)]^3}. \quad (1.6)$$

Passing to new variables

$$p \to p + k(1-z), \quad (1.7)$$

we obtain

$$\Sigma_\Lambda(k^2) = \frac{2ik_\nu}{(2\pi)^4} \int_0^{\Lambda^2} d\lambda \int_0^1 dz 2z \times$$

$$\times \int \frac{dp\,(kz - p)_\nu}{[p^2 + k^2(1-z)z - \lambda z + i0]^3}. \quad (1.8)$$

The integral

$$\int dp\, p_\nu f(p^2) \quad (1.9)$$

is equal to zero for reasons of symmetry.

In the remaining integral one can rotate the integration contour through 90° and introduce a new integration variable $p_0 \to ip_0$. As a result, the integral over p takes the form

$$I = -i \int \frac{dp}{(p^2 + c)^3}, \quad (1.10)$$

where integration is performed over the four-dimensional Euclidean space. Calculation of the integral (1.10) gives

$$I = -\frac{i\pi^2}{2c}. \quad (1.11)$$

As a result, we obtain the following expression for the function $\Sigma_\Lambda(k^2)$:

$$\Sigma_\Lambda(k^2) = -\frac{k^2}{16\pi^2} \int_0^{\Lambda^2} d\lambda \int_0^1 dz \frac{2z^2}{k^2(1-z)z - \lambda z}. \quad (1.12)$$

The integration over λ is performed explicitly. At $k^2 < 0$ we obtain

$$\Sigma_\Lambda(k^2) = \frac{2k^2}{16\pi^2} \int_0^1 dz \cdot z \ln \frac{k^2(1-z)z - \Lambda^2 z}{k^2(1-z)z}; \quad (k^2 < 0). \quad (1.13)$$

As $\Lambda \to \infty$ this expression, as might be expected, diverges logarithmically. The renormalization procedure consists in replacing the integral in (1.13) by an expression obtained by the subtraction from this integral of one or more leading terms of the expansion in a Taylor series (in this case one term is subtracted). Expanding the integrand about the point $k^2 = \varkappa$, we obtain

$$\Sigma_\Lambda(k^2) = \frac{2k^2}{16\pi^2}\left\{\int_0^1 z\,dz \ln\frac{\varkappa(1-z)z - \Lambda^2 z}{\varkappa(1-z)z} + \right.$$

$$\left. + \int_0^1 z\,dz\left[\ln\frac{k^2(1-z)z - \Lambda^2 z}{\varkappa(1-z)z - \Lambda^2 z} - \ln\frac{k^2(1-z)z}{\varkappa(1-z)z}\right]\right\}. \quad (1.14)$$

As $\Lambda \to \infty$ the second and third terms tend to a definite limit equal to

$$\Sigma_R(k^2) = -\frac{k^2}{16\pi^2}\ln\frac{k^2}{\varkappa}. \quad (1.15)$$

The first term has no limit as $\Lambda \to \infty$, and it behaves as

$$-k^2 g^{-2}(1-\tilde{z}_2) = \frac{k^2}{16\pi^2}\left(\ln\frac{\Lambda^2}{-\varkappa} + \ldots\right). \quad (1.16)$$

The separation (1.14) is not the only one possible. Choosing as the central point of the expansion a point other than $\varkappa$, we would obtain for $\Sigma_R(k^2)$ an expression differing from (1.15) by a finite polynomial in k^2. Thus, the general expression for the renormalized Green function to second order in g^2 has the form

$$-\frac{1}{k^2 + i0}\left(1 + \bar{b}_2 - \frac{g^2}{16\pi^2}\ln\frac{k^2}{\varkappa}\right), \quad (1.17)$$

where $\bar{b}_2$ is an arbitrary constant.

Substitution of the renormalized expression (1.15) for the divergent integral (1.2) is equivalent to the redefinition of the original Lagrangian. Indeed, let us replace the Lagrangian for the fictitious particles by the following expression:

$$-\frac{\mathrm{tr}}{2}\bar{c}\partial_\mu \nabla_\mu c \to -\frac{\mathrm{tr}}{2}\{\bar{c}\,\Box c - g\bar{c}\partial_\mu[A_\mu, c] +$$

$$+ (\tilde{z}_2 - 1)\bar{c}\,\Box c\}, \quad (1.18)$$

where $\tilde{z}_2$ is defined by the formula (1.16). Since the last term $\sim g^2$, we shall attribute it to the interaction Lagrangian. Then in the perturbation-theory expansion there will appear, besides the diagram in

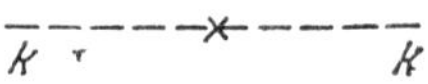

k k

Fig. 2.

Figure 1, a new diagram (Figure 2), where the cross indicates the vertex responsible for the "counterterm" $(\bar{z}_2 - 1)\bar{c}\Box c$.

Obviously the correction to the Green function corresponding to the sum of the diagrams in Figures 1 and 2 is given by the formula (1.17) (at $\tilde{b}_2 = 0$). This simple example shows that the subtraction of the leading terms of the expansion in the Taylor series is equivalent to a change (renormalization) of the original Lagrangian parameters (in this case of the normalization constant of the fictitious-particle wave function).

Let us illustrate this observation by one more example. We shall calculate the third-order correction to the vertex function $\Gamma_{\bar{c}c\mathscr{A}_\mu}$ responsible for the transition of two fictitious particles into one vector particle. The diagrams, contributing to $\Gamma_{\bar{c}c\mathscr{A}_\mu}$ in the third order in g, are presented in Figure 3.

For simplicity we shall restrict ourselves to the case of zero momentum transfer $q = 0$.

The integral

$$I_a = -ig^3\varepsilon^{abc} \int \frac{dp\,(k-p)_\alpha\,(k-p)_\mu\,k_\beta g_{\alpha\beta}}{(2\pi)^4\,(p^2+i0)\,[(k-p)^2+i0]^2} \tag{1.19}$$

corresponds to the diagram (a). Introducing intermediate regularization with the formula (1.3) and using the relation

$$\frac{1}{a^2b^2} = \int_0^1 \frac{6z\,(1-z)\,dz}{[az+b\,(1-z)]^4}, \tag{1.20}$$

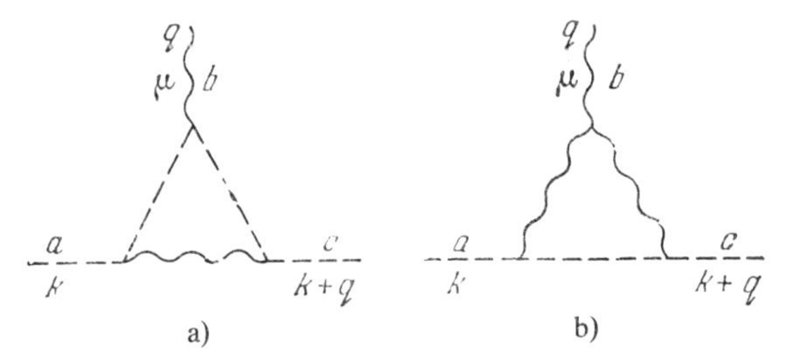

Fig. 3. Third-order corrections to the vertex function $\Gamma_{\bar{c}c\mathscr{A}_\mu}$.

we write this integral in the form

$$I_a = -\frac{ig^3\varepsilon^{abc}k_\alpha}{(2\pi)^4}\int_0^1 dz\int_0^{\Lambda^2} d\lambda\,\times$$

$$\times\int dp\,\frac{(k-p)_\alpha(k-p)_\mu 6z(1-z)}{\{[p-k(1-z)]^2+k^2(1-z)z-\lambda z\}^4}. \quad (1.21)$$

The change of variables (1.7) gives

$$I_a = -\frac{ig^3\varepsilon^{abc}k_\alpha}{(2\pi)^4}\int_0^1 dz\int_0^{\Lambda^2} d\lambda\,\times$$

$$\times\int dp\,\frac{(kz-p)_\alpha(kz-p)_\mu 6z(1-z)}{\{p^2+k^2(1-z)z-\lambda z\}^4}. \quad (1.22)$$

The odd powers of p do not give any contribution, for reasons of symmetry. For the same reasons

$$\int dp p_\mu p_\nu f(p^2) = \frac{1}{4}g_{\mu\nu}\int dp p^2 f(p^2). \quad (1.23)$$

Passing to the Euclidean metric and integrating over p, we obtain

$$I_a = -\frac{ig^3\varepsilon^{abc}k_\alpha}{(2\pi)^4}\times$$

$$\times\left\{\int_0^1 dz\int_0^{\Lambda^2} d\lambda z^2(1-z)k_\alpha k_\mu\frac{i\pi^2}{\{k^2(1-z)z-\lambda z\}^2}+\right.$$

$$\left.+g_{\alpha\mu}\int_0^1 dz\frac{(-i\pi)^2 2}{4}z(1-z)\int_0^{\Lambda^2} d\lambda\,\frac{1}{\{k^2(1-z)z-\lambda z\}}\right\}. (1.24)$$

Integration over λ gives

$$I_a = \frac{g^3}{16\pi^2}\varepsilon^{abc}k^2k_\mu\times$$

$$\times\int_0^1\left\{\frac{1}{k^2(1-z)z-\Lambda^2 z}-\frac{1}{k^2(1-z)z}\right\}z^2(1-z)\,dz+$$

$$+\frac{g^3}{32\pi^2}\varepsilon^{abc}k_\mu\int_0^1\ln\frac{k^2(1-z)z-\Lambda^2 z}{k^2(1-z)z}(1-z)\,dz \qquad (k^2<0).$$

$$(1.25)$$

The first term in the formula (1.25) tends to a definite limit as $\Lambda \to \infty$, and the second one diverges logarithmically. As in the case of the second-order diagram, the expression obtained by subtraction of the first term of the expansion in the Taylor series from the second integral tends to a definite limit. As $\Lambda \to \infty$, I_a can be expressed as

$$I_a = \frac{g^3 \varepsilon^{abc} k_\mu}{64\pi^2}\left(b^1_\varkappa - \ln\frac{k^2}{\varkappa}\right) + \frac{g^3 \varepsilon^{abc} k_\mu}{64\pi^2} \ln\frac{\Lambda^2}{-\varkappa}, \qquad (1.26)$$

where $b^1_\varkappa$ is a finite constant, depending on the choice of the point $\varkappa$.

The diagram b is calculated in an entirely analogous manner. The corresponding integral has the form

$$I_b = -ig^3\varepsilon^{abc} \times$$
$$\times \int \frac{dp}{(2\pi)^4} \frac{(k-p)_\alpha k_\beta \{2p_\mu g_{\sigma\rho} - p_\sigma g_{\mu\rho} - p_\rho g_{\mu\sigma}\} g_{\alpha\sigma} g_{\rho\beta}}{[(k-p)^2 + i0][p^2 + i0]^2}. \qquad (1.27)$$

Repeating the computations given above, we obtain as $\Lambda \to \infty$

$$I_b = \frac{g^3 \varepsilon^{abc} k_\mu}{64\pi^2}\left(b^2_\varkappa - 3\ln\frac{k^2}{\varkappa}\right) + \frac{3g^3 \varepsilon^{abc} k_\mu}{64\pi^2} \ln\frac{\Lambda^2}{-\varkappa}. \qquad (1.28)$$

Subtracting the terms proportional to $\ln[\Lambda^2/(-\varkappa)]$ from the sum $I_a + I_b$, we obtain the expression for the renormalized vertex function in the form

$$I^R = \frac{g^3 \varepsilon^{abc} k_\mu}{16\pi^2}\left(\tilde{b}_1 - \ln\frac{k^2}{\varkappa}\right), \qquad (1.29)$$

where $\tilde{b}_1$ is an arbitrary constant. The subtraction performed is equivalent to the insertion in the Lagrangian of the counterterm

$$\frac{\mathrm{tr}}{2}\{g(\tilde{z}_1 - 1)\bar{c}\partial_\mu[\mathcal{A}_\mu, c]\}, \qquad (1.30)$$

where

$$\tilde{z}_1 - 1 = -\frac{g^2}{16\pi^2}\left(\ln\frac{\Lambda^2}{-\varkappa} + \tilde{b}_1\right). \qquad (1.31)$$

In conclusion we shall give without computation the expressions for the lowest-order corrections to the Green function of the Yang–Mills field and for the three-point vertex Γ_A, which are described by the diagrams presented in Figures 4 and 5.

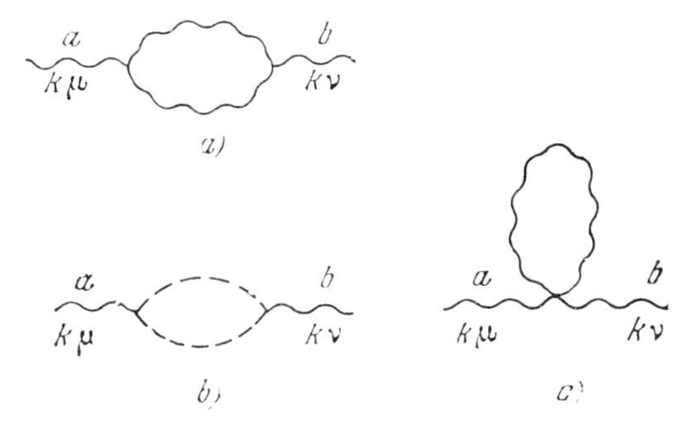

Fig. 4. Second-order corrections to the Green function of the Yang–Mills field.

The counterterms which remove the divergences from these diagrams have the form

$$-\frac{\mathrm{tr}}{8}\{(z_2-1)(\partial_\nu\mathcal{A}_\mu-\partial_\mu\mathcal{A}_\nu)^2+2g(z_1-1)\times \\ \times(\partial_\nu\mathcal{A}_\mu-\partial_\mu\mathcal{A}_\nu)[\mathcal{A}_\mu,\mathcal{A}_\nu]\}, \quad (1.32)$$

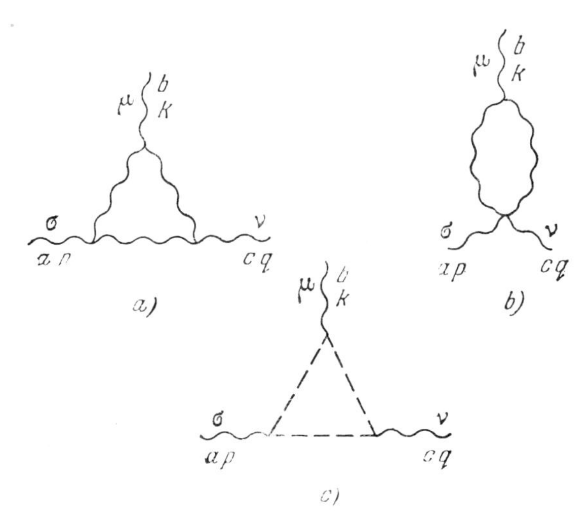

Fig. 5. Third-order corrections to the vertex function Γ_{A^3}.

where

$$z_2 - 1 = \frac{g^2 \cdot 5}{24\pi^2} \ln \frac{\Lambda^2}{-\varkappa} + b_2, \tag{1.33}$$

$$z_1 - 1 = \frac{g^3}{12\pi^2} \ln \frac{\Lambda^2}{-\varkappa} + b_1, \tag{1.34}$$

in which b_1 and b_2 are arbitrary finite constants.

As is seen, for the removal of divergences from the diagrams under consideration it is indeed sufficient to redefine the parameters of the original Lagrangian. Note, however, that the counterterms (1.16), (1.30), (1.32) are not, generally speaking, gauge-invariant. Their explicit form depends on the intermediate regularization used and the choice of the subtraction points. In particular, to remove the divergences from the two-point Green function it may be necessary to introduce a manifestly noninvariant counterterm.

$$\delta m \operatorname{tr} \{\mathcal{A}_\mu \mathcal{A}_\mu\}. \tag{1.35}$$

Of course, it is the renormalized finite matrix elements and not the counterterms themselves that have physical meaning. For the theory to remain self-consistent it is necessary that the renormalized matrix elements satisfy the relativity principle. This requirement entails the specific features of the renormalization procedure in gauge theories, which are to be investigated in the present chapter.

4.2. THE R-OPERATION AND COUNTERTERMS

In the preceding section we have discussed the procedure of removing divergences from the simplest diagrams. The considered examples contain only one integration over dk, and so, when the intermediate regularization is removed, in order to provide for the corresponding functions to tend to a definite limit, it is sufficient to subtract one or more leading terms of their expansion in a Taylor series in the external momenta. As we have seen, such a subtraction is equivalent to the redefinition of the original Lagrangian, that is, to the introduction of counterterms.

To more complicated diagrams such as, for example, the one presented in Figure 6, there correspond integrals of the form

$$\int f(p_1, \ldots, p_m, k_1, \ldots, k_n)\, dk_1 \ldots dk_n, \tag{2.1}$$

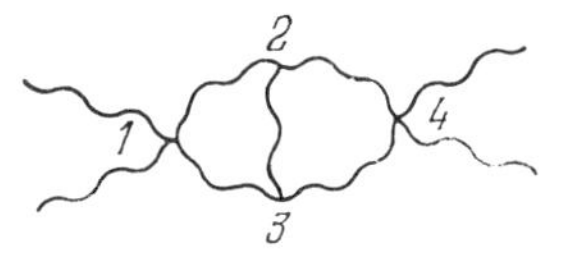

Fig. 6.

which may diverge not only when all k_i tend to infinity simultaneously, but also when some of the arguments k_i tend to infinity while the rest remain fixed. In this case it is said that the diagram has diverging subgraphs. For the diagrams in Figure 6 such subgraphs are represented by the combinations of vertices (1,2,3) and the lines connecting them, and by the combination of vertices (2,3,4) and the lines connecting them. For diagrams with divergent subgraphs, the simple recipe for the removal of divergences formulated in Section 4.1 is already insufficient. In this case the problem is solved by means of the *R*-operation of Bogolubov and Parasyuk, which for any Feynman diagram provides a corresponding finite coefficient function. A detailed discussion of the *R*-operation may be found in the book of N. N. Bogolubov and D. V. Shirkov [1], and we shall not repeat it here. For our aims it is sufficient to know that the *R*-operation is equivalent to the insertion in the Lagrangian of counterterms, which may be represented as series in the coupling constant. In order to formulate the corresponding recipe we shall need several definitions. A diagram is said to be connected if it cannot be separated into parts which are not connected to each other by lines. A diagram is called strongly connected or one-particle-irreducible if it cannot be transformed into disconnected diagrams by the removal of a single line. A strongly connected diagram with all external lines cut off will be called a proper vertex function and denoted as $\Gamma(x_1,\ldots,x_n)$. A strongly connected Green function $G(x_1,\ldots,x_n)$ is expressed in terms of the proper vertex function $\Gamma(x_1,\ldots,x_n)$ by the relation

$$G(x_1, \ldots, x_n) = \\ = \int dx_1' \ldots dx_n' G_1(x_1 - x_1') \ldots G_n(x_n - x_n') \Gamma(x_1', \ldots, x_n'), \tag{2.2}$$

where $G_i(x_i - x_i')$ is the two-particle Green function corresponding to the ith external line. The topological structure of the diagrams is conveniently characterized by the number of independent cycles contained in the given diagram. Diagrams with one cycle are called one-loop diagrams; with two cycles, two-loop diagrams; etc. Diagrams with a given number of loops are terms of the same order in the quasiclassical expansion in Planck's constant $\hbar$ of the S-matrix or of the generating functional for the Green functions. For this reason they form an invariant combination; that is, all the symmetry properties of the complete S-matrix are satisfied independently for a combination of diagrams with a fixed number of loops.

To characterize the procedure for the removal of divergences, the notion of a diagram index is introduced. Let the coefficient function with the Fourier transform

$$J(k) = \int \prod_{1 \leqslant q \leqslant n} \delta\left(\sum p - k_q\right) \prod_{l=1}^{L} D_l(p_l)\, dp_l, \tag{2.3}$$

where index q numbers vertices and l numbers the internal lines correspond to a strongly connected diagram. In the argument of the δ-function there is the algebraic sum of momenta which enter into the vertex number q. The Green function $D_l(p_l)$ has the form

$$D_l(p_l) = Z(p_l)\left(m_l^2 - p_l^2\right)^{-1}, \tag{2.4}$$

where $Z(p_l)$ is a polynomial of degree r_l.

Let us perform a scale transformation of all the momentum variables (and masses): p_i, $k_i \rightarrow ap_i$, ak_i. If the integral $J(k)$ converges, then under such a transformation it will be multiplied by a^{ω}, where the diagram index ω consists of the following factors: each internal line contributes $r_l - 2$, resulting in $\sum_{l=1}^{L} (r_l - 2)$, where L is the total number of internal lines. In the formula (2.3) integration is performed over L variables p_l; however, $n - 1$ integrals are removed by δ-functions (one δ-function expresses the conservation law for the total momentum). Therefore there remain $4(L - n + 1)$ independent differentials, which give a total contribution of $4(L - n + 1)$. If the interaction Lagrangian contains derivatives, then each vertex with m derivatives introduces an additional factor m. Summing these factors, we obtain

$$\omega = \sum_l (r_l + 2) - 4(n - 1) + mn. \tag{2.5}$$

The index ω defines the degree of growth of the coefficient function when a uniform extension of all the momentum variables takes place. At $\omega \geq 0$ this definition, in general, loses its sense, since the corresponding integral diverges. In this case the diagram index defines the superficial degree of growth. From the nonnegativeness of the diagram index follows the divergence of the corresponding integral. The opposite is not true in general, since the index of a diagram characterizes its behavior only when a simultaneous extension of all momenta occurs, and has nothing to do with its behavior when some of the integration variables tend to infinity while the rest remain fixed. In other words, a diagram with a negative index may have divergent subgraphs. Negativeness of the index is a sufficient condition for the convergence of primitively divergent diagrams, that is, of diagrams which become convergent if any internal line is broken. The diagrams considered in the first section are primitively divergent, whereas the diagram in Figure 6 has divergent subgraphs (1,2,3) and (2,3,4). Obviously one-loop diagrams can only be primitively divergent.

We can now formulate a recipe for the removal of divergences from arbitrary diagrams by the insertion of counterterms. First of all we shall introduce an intermediate regularization making all the integrals convergent [for instance, by means of the formula (1.3)].

First, let us consider one-loop diagrams. As we have already seen, for the corresponding coefficient functions to tend to a definite limit when the regularization is removed, it is sufficient to subtract from them several leading terms of the expansion in the Taylor series in the external momenta. Such a subtraction is, in turn, equivalent to the insertion in the Lagrangian of counterterms, that is, to the substitution of $\mathscr{L}_\Lambda + \triangle \mathscr{L}_1$ for the original regularized Lagrangian $\mathscr{L}_\Lambda$.

The explicit expression for the counterterms $\triangle \mathscr{L}_1$ is constructed in the following way. Let G^n_s be a strongly connected diagram with n vertices and s external lines A_μ, having a nonnegative index ω. To it there corresponds the proper vertex function $\Gamma^n_s(x_1,\ldots,x_n)$. The subtracted polynomial consists of the first terms in the expansion of the Fourier transform of Γ^n_s in the Taylor series, and in the coordinate representation it has the form

$$Z_{(\mu}\left(\frac{\partial}{\partial x_i}\right)\delta(x_1 - x_2)\ \ldots\ \delta(x_{n-1} - x_n), \tag{2.6}$$

where Z is a symmetric polynomial of the order ω. In order to obtain the counterterm corresponding to the given diagram, it is necessary to multiply the expression (2.6) by the product

$$\mathscr{A}_{\mu_1}(x_1) \ \ldots \ \mathscr{A}_{\mu_s}(x_s), \tag{2.7}$$

and then to sum the obtained expression over $\mu_1,\ldots,\mu_s$ and integrate over all variables $x_1,\ldots,x_n$ except one. (If besides vector external lines, the diagram contains others—for instance, spinor and scalar lines—then all the reasoning remains the same except that symmetrization is applied only to lines of the same type.)

We shall now construct two-loop diagrams using as a Lagrangian $\mathscr{L}_\Lambda + \triangle \mathscr{L}_1$. The two-loop diagrams thus constructed do not now contain divergent subgraphs; that is, when the intermediate regularization is removed, divergence appears only with the simultaneous tending of all integration variables to infinity. This fact is quite obvious if the divergent subgraphs do not overlap, as happens, for example, with the diagram presented in Figure 7. In this case the counterterm $\triangle \mathscr{L}_1$, removing the divergence from the subgraph (2.3), has the form

$$-\frac{(z_2-1)}{4}(\partial_\mu A_\nu^a - \partial_\nu A_\mu^a)^2, \tag{2.8}$$

and the Lagrangian $\mathscr{L} + \triangle \mathscr{L}_1$ generates, in addition to the diagram in Figure 7, also the one presented in Figure 8, where the cross indicates the vertex (2.8). The integral corresponding to the sum of the diagrams in Figures 7 and 8 diverges only when all momenta tend simultaneously to infinity, and for the removal of the divergence it is again sufficient to subtract from the integral the first two terms of the expansion in the Taylor series, which is equivalent to the insertion in the Lagrangian of a new counterterm $\triangle \mathscr{L}_2$:

$$\mathscr{L} \to \mathscr{L} + \Delta\mathscr{L}_1 + \Delta\mathscr{L}_2. \tag{2.9}$$

The proof of an analogous statement when overlapping divergent subgraphs are present (for instance, as in Figurc 6) is morc complicated, and we shall not present it here.

Proceeding in this manner, we come to the renormalized

Fig. 7.

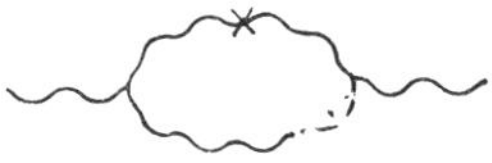

Fig. 8.

Lagrangian

$$\mathscr{L}_R = \mathscr{L} + \Delta\mathscr{L}_1 + \ldots + \Delta\mathscr{L}_n, \tag{2.10}$$

where $\Delta\mathscr{L}_i$ are local polynomials in the fields and their derivatives, for which all diagrams containing not more than n loops converge to a finite limit when the intermediate regularization is removed. Obviously, in the framework of perturbation theory, we thus calculate the Green functions to any finite order n. With increasing n the total number of counterterms, of course, increases; however, the number of counterterms of various types may turn out to remain finite. (We call the functional dependence of a counterterm on the fields the *type* of the counterterm.) In this case it is said that the theory is renormalized. A renormalizable theory is determined by a finite number of parameters, having the meaning of physical charges and masses. But if the number of the counterterm types increases infinitely, (that is, in the higher orders of perturbation theory there appear structures containing more and more fields and their derivatives), then the theory is called nonrenormalizable. Since the insertion of a new counterterm is equivalent to the appearance of a new arbitrary constant (the position of the point of subtraction), nonrenormalizable theories are not determined by a finite number of parameters. For nonrenormalizable Lagrangians the perturbation-theory method seems to be useless, and we shall not consider them here.

The explicit form of the counterterms depends on the concrete intermediate regularization and on the choice of the subtraction point, that is, the center of expansion in the Taylor series. An inconvenient choice of regularization may render the analysis of the renormalized theory extremely difficult. In the case of gauge theories the so-called invariant regularizations, conserving the formal symmetry properties of the nonrenormalized theory, are especially convenient.

4.3 INVARIANT REGULARIZATIONS. THE PAULI–VILLARS PROCEDURE

The counterterm form of the R-operation is convenient for the investigation of Yang–Mills fields, since it allows us to take into account symmetry properties in a simple and explicit manner. As we have already seen in the previous chapter, the relativity principle allows us to construct a perturbation theory for Yang–Mills fields, proceeding from various gauges. The gauges in which the S-matrix is formally unitary (the Coulomb or Hamilton gauges for the massless Yang–Mills field, the unitary gauge for the theory with spontaneously broken symmetry) are inconvenient from the point of view of the renormalization procedure. In the first two cases there is no manifest relativistic invariance, and in the latter explicit renormalizability is absent. Significantly more convenient in this sense are the manifestly covariant gauges, such as the Lorentz one, for which, as we shall see, renormalizability is obvious. However, in the Lorentz gauge we cannot construct a Hamiltonian formulation of the theory, and therefore the unitarity of the S-matrix is not obvious. From the viewpoint of the operator formalism the S-matrix in the Lorentz gauge acts in the "big" space containing both physical and nonphysical states (longitudinal and time "photons", scalar fermions, Goldstone bosons) and, generally speaking, is unitary only in this space, in which the metric is indefinite. The unitarity of the S-matrix in the physical subspace, the states of which correspond to fields of matter and to transverse vector quanta, is a consequence of the relativity principle, according to which all observables are independent of the gauge condition actually chosen. This is confirmed by the explicit calculations of the previous chapter, where it was shown that an explicitly unitary generating functional for the coefficient functions of the S-matrix in the Coulomb gauge may be transformed identically into a functional corresponding to the Lorentz gauge. The reasoning given, however, was of a formal character, since we did not pay attention to the divergences appearing in perturbation-theory calculations of these functionals. Indeed, in quantum theory the relativity principle should be applied to renormalized entities free of divergences. The transfer of this principle to a renormalized theory is not trivial. Renormalization is equivalent to the redefinition of the original Lagrangian. Therefore it is necessary to prove that the renormalized Lagrangian is gauge-invariant. Then we may apply to it the reasoning of the preceding chapter and rigorously prove the equivalence of various gauges and, consequently, the unitarity of the S-matrix.

This latter statement needs to be clarified. As we have already seen, the explicit form of the renormalized Lagrangian depends on the intermediate regularization used. What we have said above applies only to the invariant intermediate regularization, that is, to the regularization which conserves the formal symmetry properties of the nonrenormalized theory. This, of course, does not mean that for the calculation of the S-matrix one is not to use a noninvariant regularization. In that case, however, the regularized theory is gauge-noninvariant, and the reasoning of the previous chapter, which demonstrates the equivalence of various gauges, is not applicable to it. The relativity principle is now valid only for the renormalized S-matrix, when the regularization is removed. All this complicates the proof of unitarity and makes it less clear. Therefore we shall start describing the renormalization procedure for the Yang–Mills theory by constructing a gauge-invariant intermediate regularization. The specific features of the invariant regularization of gauge theories are due to the interaction of Yang–Mills fields in vacuum. The interaction with fields of matter does not introduce any difficulties: the corresponding diagrams are regularized by means of the obvious generalization of the gauge-invariant Pauli–Villars procedure.

We shall show this by using the example of the interaction of the Yang–Mills field with a spinor field ψ described by the Lagrangian (1.3.1). The generating functional for the Green functions has the form

$$Z(J_\mu, \bar{\eta}, \eta) = N^{-1} \int \exp\Big\{ i \Big[\mathscr{L}_{YM} + \frac{1}{2\alpha} (\partial_\mu A_\mu^a)^2 + \\ + i\bar{\psi}(\hat{\partial} - g\Gamma^a \hat{A}^a)\psi - \mu_0 \bar{\psi}\psi + J_\mu^a A_\mu^a + \\ + \bar{\eta}\psi + \bar{\psi}\eta \Big] dx \Big\} \det M \prod_x dA_\mu \, d\bar{\psi} \, d\psi. \tag{3.1}$$

Divergent diagrams, not containing internal vector lines (spinor cycles) are regularized in the same manner as in electrodynamics, that is, by subtracting analogous cycles along which the spinor fields with masses μ_i propagate. Actually, if we are interested only in spinor cycles, we may set the sources $\bar{\eta}$, η equal to zero. The remaining Gaussian integral over ψ and $\bar{\psi}$ is calculated explicitly. It is equal to $\det X_0$, where

$$X_0 = i\gamma_\mu \partial_\mu - \mu_0 - ig\Gamma^a \gamma_\mu A_\mu^a. \tag{3.2}$$

The regularization consists in the substitution for $\det X_0$ of the product

$$\det X_0 \to \det X_0 \prod_{j=1}^{n} (\det X_j)^{c_j} = \\ = \exp\Big\{ \operatorname{Tr} \ln X_0 + \sum_{j=1}^{n} c_j \operatorname{Tr} \ln X_j \Big\}, \tag{3.3}$$

where the operators X_j are constructed analogously to the operator X_0:

$$X_j = i\hat{\partial} - \mu_j - ig\Gamma^a \hat{A}^a, \tag{3.4}$$

and the coefficients c_j satisfy the conditions

$$\sum c_j + 1 = 0, \quad \sum c_j \mu_j^2 = 0. \tag{3.5}$$

In order to verify that the substitution (3.3) really regularizes the spinor cycles, let us represent det X_j in the form

$$\det X_j = \det (i\hat{\partial} - \mu_j) \det \{1 - g(i\hat{\partial} - \mu_j)^{-1} i\Gamma^a \hat{A}^a\}. \tag{3.6}$$

The first factor does not depend on the fields A_μ and therefore can be included in the normalization constant N. The second factor may be transformed into

$$\begin{aligned} &\exp \{\mathrm{Tr} \ln \{1 - g(i\hat{\partial} - \mu_j)^{-1} i\Gamma^a \hat{A}^a\}\} = \\ &= \exp \Big\{ - \Big[\frac{(ig)^2}{2} \mathrm{tr} \int \Big[\Gamma^{a_1} \hat{A}^{a_1}(x_1) S^j(x_1 - x_2) \times \\ &\times \Gamma^{a_2} \hat{A}^{a_2}(x_2) S^j(x_2 - x_1) \Big] dx_1\, dx_2 + \ldots \\ &\ldots + \frac{(ig)^n}{n} \mathrm{tr} \int \Gamma^{a_1} \hat{A}^{a_1}(x_1) S^j(x_1 - x_2) \ldots \\ &\ldots \Gamma^{a_n} \hat{A}^{a_n}(x_n) S^j(x_n - x_1)\, dx_1 \ldots dx_n \Big] \Big\}, \end{aligned} \tag{3.7}$$

where $S^j(x)$ is the spinor Green function

$$S^j(x) \equiv (i\hat{\partial} - \mu_j)^{-1} = \frac{-1}{(2\pi)^4} \int \frac{\mu_i + \hat{p}}{\mu_j^2 - p^2 - i0} e^{-ipx}\, dp. \tag{3.8}$$

Passing to Fourier transforms, the nth term in the exponential can be written as

$$\begin{aligned} &\mathrm{const} \cdot \int \Big[\int dp \frac{\mathrm{tr}\,[\gamma_{\nu_1}(\mu_j + \hat{p}) \ldots \gamma_{\nu_n}(\mu_j + \hat{p} + k_{n-1})]}{(\mu_j^2 - p^2)(\mu_j^2 - (p + k_1)^2) \ldots (\mu_j - (p + k_{n-1})^2)} \times \\ &\times \mathrm{tr} \big[\Gamma^{a_1} A^{a_1}_{\nu_1}(k_1) \ldots \Gamma^{a_n} A^{a_n}_{\nu_n}(k_n) \big] \times \\ &\times \delta(k_n - k_1 - \ldots - k_{n-1}) \Big] dk_1 \ldots dk_n. \end{aligned} \tag{3.9}$$

Here the first trace is related to the spinor indices and the second one to the internal degrees of freedom. At $n \leq 4$ the integral over p diverges. At large p the integrand in this integral may be represented as a series in μ_i,

$$\frac{P_n(p) + \mu_j^2 P_{n-2}(p) + \ldots + \mu_j^n}{P_{2n}(p) + \mu_j^2 P_{2n-2}(p) + \ldots + \mu_j^{2n}} = \frac{P_n(p)}{P_{2n}(p)} + \frac{P_n(p)}{P_{2n}(p)} \times$$
$$\times \left[\frac{P_{n-2}(p)}{P_n(p)} - \frac{P_{2n-2}(p)}{P_{2n}(p)} \right] \mu_j^2 + \ldots, \tag{3.10}$$

where $P_j(p)$ is a polynomial of the order j in p. The coefficient of μ_j^{2k} at large p behaves as p^{-n-2k}. If the coefficients c_j satisfy the conditions (3.5), then the two highest-order terms in the asymptotic expansion in p of the integrand in the sum (3.3) fall out, and the asymptotic behavior of the regularized expression is p^{-n-4}. Thus, all the integrals over p converge.

The regularized generating functional can be represented by a path integral in which the exponent contains the local action, if one uses the representation of det X_j in the form

$$[\det X_j]^{\pm 1} =$$
$$= \int \exp\left\{ i \int [i\bar{\psi}_j(\hat{\partial} - g\Gamma^a \hat{A}^a)\psi_j - \mu_j \bar{\psi}_j \psi_j]\, dx \right\} \prod_x d\bar{\psi}_j\, d\psi_j, \tag{3.11}$$

where $\bar{\psi}_j$, ψ_j are the auxiliary spinor variables. The exponent of the determinant on the left-hand side of this equality depends on the commutation properties of the fields ψ_j. The exponent $+1$ corresponds to anticommuting variables, and -1 corresponds to commuting variables. By choosing integers for the coefficients c_j in the formula (3.3), we can represent the regularizing factor $\prod_{j=1}^{n} (\det X_j)^{c_j}$ in the form

$$\prod_{j=1}^{n} (\det X_j)^{c_j} = \int \exp\left\{ i \int \sum_{j=1}^{n} \left[\sum_{k=1}^{|c_j|} [i\bar{\psi}_{jk}(\hat{\partial} - g\Gamma^a \hat{A}^a)\psi_{jk} - \right.\right.$$
$$\left.\left. - \mu_j \bar{\psi}_{jk} \psi_{jk}] \right] dx \prod_{x,j,k} d\bar{\psi}_{jk}\, d\psi_{jk} \right\}. \tag{3.12}$$

Here the coefficients c_j and the masses μ_j are assumed to satisfy the conditions (3.5). The commuting auxiliary fields $\bar{\psi}_{jk}$, ψ_{jk} correspond to the negative coefficents c_j, and the anticommuting auxiliary fields correspond to the positive coefficients c_j.

Adding to the action in the exponent (3.1) the action from the right-hand side of the equality (3.12), we obtain the regularized generating functional in the form of a path integral of exp{$i \times$ local action}. The action (3.12) is manifestly invariant with respect to simultaneous gauge transformations of the fields $\mathscr{A}_\mu$, $\bar{\psi}_{jk}$, ψ_{jk}, and therefore the regularization (3.3) does not change the symmetry properties of the generating functional.

The generalization of this procedure to the case when the Yang–Mills field interacts also with a scalar field is obvious. The only difference is in that, since scalar fields are commuting entities, the sum of the closed cycles is equal to $(\det Y_0)^{-1}$ and the regularization consists in the substitution

$$(\det Y_0)^{-1} \to (\det Y_0)^{-1} \prod_j (\det Y_j)^{-c_j}. \tag{3.13}$$

The Pauli–Villars regularization is applicable in those cases when the interaction Lagrangian is quadratic in the fields forming divergent cycles. Therefore, it cannot be generalized to the Yang–Mills field itself. Here one has to resort to more sophisticated methods. We shall further restrict ourselves to the consideration of the Yang–Mills field in vacuum.

At present there exist two methods for invariant regularization of non-Abelian gauge theories: the method of higher covariant derivatives and the method of dimensional regularization.

The first method is, actually, an invariant generalization of the standard regularization procedure, when free propagators are regularized by the subtraction

$$-\frac{1}{k^2} \to -\frac{1}{k^2} - \frac{1}{\Lambda^2 - k^2} = -\frac{1}{k^2 - \Lambda^{-2}k^4} \tag{3.14}$$

(for simplicity the scalar propagator is written). Such a subtraction is equivalent to the insertion in the Lagrangian of terms with higher derivatives:

$$\frac{1}{2}\partial_\mu\varphi\partial_\mu\varphi \to \frac{1}{2}\partial_\mu\varphi\partial_\mu\varphi + \frac{1}{2\Lambda^2}\Box\varphi\,\Box\varphi. \tag{3.15}$$

In the case of the Yang–Mills field such a procedure violates the gauge invariance, since an ordinary derivative is not a covariant object. A natural generalization of the regularization (3.15) consists in adding to the Yang–Mills Lagrangian a term containing higher covariant derivatives, for example,

$$\mathcal{L}_{YM} \to \mathcal{L}_{YM}^{\Lambda} = \frac{1}{8} \operatorname{tr} \left\{ \mathcal{F}_{\mu\nu}\mathcal{F}_{\mu\nu} + \frac{1}{\Lambda^2} \nabla_\alpha \mathcal{F}_{\mu\nu} \nabla_\alpha \mathcal{F}_{\mu\nu} \right\} =$$

$$= \frac{1}{8} \operatorname{tr} \left\{ \mathcal{F}_{\mu\nu}\mathcal{F}_{\mu\nu} + \frac{1}{\Lambda^2} (\partial_\alpha \mathcal{F}_{\mu\nu} - g [\mathcal{A}_\alpha, \mathcal{F}_{\mu\nu}])^2 \right\}. \quad (3.16)$$

The substitution (3.16) leads to the desired modification of the free propagator. However, the cost of achieving invariance is the appearance of new vertices in the interaction Lagrangian. Below we shall discuss this regularization in more detail, but now we just point out that due to the appearance of new vertices with derivatives, the regularization is only partial—in the regularized theory the second-, third-, and fourth-order diagrams remain divergent. Thus the method of higher derivatives alone does not solve the problem completely, but only reduces the problem to the investigation of a superrenormalizable theory, that is, a theory generating a finite number of divergent diagrams. Below it will be shown that the remaining diagrams can be regularized by means of a somewhat modified Pauli–Villars procedure. As a result we shall describe an explicitly invariant Lagrangian which generates convergent (for finite regularization parameters) Feynman diagrams. The defect of this method is that it is relatively cumbersome. Due to the appearance of new vertices in the interaction Lagrangian, the number of diagrams is greatly increased, impeding practical calculations. However, for investigating fundamental problems of unitarity and renormalization this method is the most convenient one, because the existence of a manifestly invariant expression for the regularized action allows us automatically to apply to the regularized case the reasoning of the previous chapter on the equivalence of various gauges, and thus to prove the unitarity of the renormalized theory.

Unlike the method of higher covariant derivatives, dimensional regularization is not reduced to some modification of the original Lagrangian, but deals directly with the Feynman diagrams. This method is based on two observations:

1. The formal symmetry relations between Green functions (generalized Ward identities) do not depend on the dimensionality of the space time (n).
2. At sufficiently small or complex n all diagrams correspond to convergent integrals.

Thus, generalized Ward identities can be proven rigorously in the region of n where all integrals converge, and then by analytical continuation one can pass over to $n = 4$.

The method of dimensional regularization has turned out to be convenient for the calculation of concrete diagrams and is quite widely used in practical calculations. It has, however, some shortcomings from the viewpoint of the investigation of matters of principle.

Since if n is noninteger or complex, no Lagrangian can be found to correspond to the regularized theory, the simple proof of unitarity based on the change of variables in the path integral is not applicable, and it is then necessary to deal directly with Feynman diagrams, which is significantly more laborious. Additional difficulties arise for the regularization of theories containing fermions. Since the algebra of the γ-matrices depends crucially on the dimensionality of the space, such theories require special consideration.

Thus, the method of higher covariant derivatives and the method of dimensional regularization, in a sense, complement each other; the first method is more convenient for general proofs which require, in fact, only the existence of the invariant regularized action, while the second method is more effective for calculating concrete processes.

4.4 THE METHOD OF HIGHER COVARIANT DERIVATIVES

The regularization will include two steps: first, by inserting in the Lagrangian higher covariant derivatives we shall pass to the superrenormalizable theory, in which only a finite number of one-loop diagrams are involved, and then we shall regularize the one-loop diagrams, using the modified Pauli–Villars procedure.

The modification of the Lagrangian (3.16) is insufficient to provide for the convergence of all the diagrams containing more than one loop. The Lagrangian (3.16), although corresponding to the superrenormalizable theory, generates a divergent two-loop self-energy diagram of the fourth order. In order to remove this divergence also, we insert in the Lagrangian a term containing fourth-order covariant derivatives:

$$\mathscr{L}_{YM} \to \mathscr{L}_{\Lambda} = \frac{1}{8} \operatorname{tr} \left\{ \mathscr{F}_{\mu\nu}\mathscr{F}_{\mu\nu} + \frac{1}{\Lambda^4} \nabla^2\mathscr{F}_{\mu\nu}\nabla^2\mathscr{F}_{\mu\nu} \right\}. \quad (4.1)$$

The gauge invariance of the regularized Lagrangian is obvious.

The regularized generating functional for the Green functions has the form

$$Z_\Lambda(J_\mu) = N^{-1}\int \exp\left\{i\int\left[\mathscr{L}_\Lambda(x) + \frac{1}{2\alpha}\{f(\Box)\,\partial_\mu A_\mu^a\}^2 + \right.\right.$$
$$\left.\left. + J_\mu^a A_\mu^a\right]dx\right\}\det M \prod_x d\mathscr{A}, \qquad (4.2)$$

where $f(\Box)$ is an arbitrary function of the d'Alambertian operator, defining the concrete form of the generalized α-gauge. The regularized free propagator of the Yang–Mills field is constructed in the usual manner:

$$D_{\mu\nu}^{ab} = \delta^{ab}\left[-\left(g_{\mu\nu} - \frac{k_\mu k_\nu}{k^2}\right)\frac{1}{k^2 + \Lambda^{-4}k^6} - \frac{\alpha k_\mu k_\nu}{k^4 f^2(-k^2)}\right]. \qquad (4.3)$$

In the Lorentz gauge ($\alpha = 0$) at large k the propagator behaves as k^{-6}. If $\alpha \neq 0$, then we shall choose the function $f(-k^2)$ to be such as not to impair the asymptotic behavior of the propagator as $k \to \infty$ and $k \to 0$, for example,

$$f(-k^2) = k^2 - \varkappa^2, \qquad (4.4)$$

where $\varkappa^2$ is an arbitrary parameter.

Explicitly writing out the term $\sim\Lambda^{-4}$ in the Lagrangian (4.1),

$$\frac{\mathrm{tr}}{8\Lambda^4}\{[\Box(\partial_\nu\mathscr{A}_\mu - \partial_\mu\mathscr{A}_\nu)]^2 -$$
$$-2[\Box(\partial_\nu\mathscr{A}_\mu - \partial_\mu\mathscr{A}_\nu)]\,\partial_\alpha[\mathscr{A}_\alpha, \mathscr{F}_{\mu\nu}] + \dots$$
$$\dots + [\mathscr{A}_\beta[\mathscr{A}_\alpha[\mathscr{A}_\mu, \mathscr{A}_\nu]]]\}^2, \qquad (4.5)$$

we see that it generates vertices with three, four, five, six, seven, and eight outgoing lines. The maximal number of derivatives in each of these vertices is 5,4,3,2,1,0, respectively. Let us now calculate the index for an arbitrary diagram. Taking into account that in our case $r_l = -4$, we obtain that the index of a diagram, containing n_k vertices with k outgoing lines, L_{in} internal lines, and L_{ex} external lines is given by the formula

$$\omega \leqslant 4 + n_3 - n_5 - 2n_6 - 3n_7 - 4n_8 - 2L_{in} =$$
$$= 6 - 2\Pi - n_3 - 2n_4 - 3n_5 - 4n_6 - 5n_7 - 6n_8, \qquad (4.6)$$

where Π is the number of closed loops.

It is not difficult to see that only integrals corresponding to one-loop second-order diagrams with two external lines, or third-order diagrams with three external lines, or fourth-order diagrams with four external lines, can be divergent. Analogous divergent diagrams are also generated by the determinant det M. Converging integrals correspond to all other diagrams, including one-loop diagrams with external lines corresponding to fictitious c-particles.

For regularization of one-loop diagrams it is natural to try to use the Pauli–Villars procedure. It is obviously sufficient to regularize the strongly connected one-loop diagrams.

The total contribution of the closed cycles with external lines A_μ may be represented as

$$Z_0 = N^{-1} \int \exp\left\{ i \int \left[\frac{1}{2} \frac{\delta^2 S_\Lambda}{\delta A^a_\mu(x)\, \delta A^b_\nu(y)} q^a_\mu(x)\, q^b_\nu(y) \right] dx\, dy + \right.$$

$$\left. + \frac{1}{2\alpha} \int \{ f(\Box)\, \partial_\mu q^a_\mu(x) \}^2 dx \right\} \det M(A) \prod_x dq_\mu . \qquad (4.7)$$

Besides strongly connected diagrams, the expansion of the functional (4.7) contains also their products, that is, disconnected diagrams. These latter ones, obviously, automatically become finite if their connected components are regularized.

We have used here as an argument the fields A_μ instead of the sources J_μ, having in mind that the one-loop diagrams generated by the functional Z_0 can be included as subgraphs in more complicated diagrams, that is, will be integrated over A_μ with a certain weight.

The expansion of $Z_0(A)$ in perturbation-theory series generates closed cycles, along which vector (q_μ) and scalar (c) particles of zero mass propagate. By analogy with the procedure performed for the cycles of fields of matter, one could regularize $Z_0(A)$ by subtracting analogous cycles along which vector and scalar particles with masses μ_j propagate. However, such a subtraction would violate gauge invariance. Unlike the generating functional for one-loop diagrams of matter fields, which is invariant with respect to gauge transformations of its arguments, the functional $Z_0(A_\mu)$ has no such property. This is due to the presence in it of a term fixing the gauge, and of cycles of fictitious particles, which violate the explicit gauge invariance. Nevertheless, as we shall now show, the divergent part of the functional $Z_0(A_\mu)$ is invariant under gauge transformations of the fields A_μ. To an accuracy of finite terms, which do not need regularization, the functional (4.7) may be transformed into the manifestly invariant form

$$Z_0 = N^{-1} \int \exp\left\{ i \int \left[\frac{1}{2} \frac{\delta^2 S_\Lambda}{\delta A^a_\mu(x)\, \delta A^b_\nu(y)} q^a_\mu(x)\, q^b_\nu(y) \right] dx\, dy \right\} \times \times \det \nabla^2_\mu \prod_x \delta(\nabla_\mu q_\mu)\, dq_\mu + \ldots, \tag{4.8}$$

where

$$\nabla_\mu q_\mu = \partial_\mu q_\mu - g[\mathscr{A}_\mu, q_\mu], \tag{4.9}$$

and $\cdots$ denotes the finite terms not to be regularized.

To demonstrate this, we shall use a method already familiar to us. We introduce the functionals $\Delta_V(\mathscr{A}_\mu, q_\mu)$ and $\Delta_W(\mathscr{A}_\mu, q_\mu)$, defined by the conditions

$$\Delta_V(\mathscr{A}_\mu, q_\mu) \int \prod_x \delta(\nabla_\mu q^\omega_\mu)\, d\omega = 1, \tag{4.10}$$

$$\Delta_W(\mathscr{A}_\mu, q_\mu) \int \prod_x \delta(\partial_\mu q^\omega_\mu - W(x))\, d\omega = 1, \tag{4.11}$$

where the gauge transformation $q_\mu \to q^\omega_\mu$ represents a shift by a function depending on $\mathscr{A}_\mu$, ω:

$$q^\omega_\mu = q_\mu + \frac{1}{\varepsilon}\{\mathscr{A}^\omega_\mu - \mathscr{A}_\mu\} = = q_\mu + \frac{1}{\varepsilon}\{\partial_\mu u - g[\mathscr{A}_\mu, u] + O(u^2)\}, \tag{4.12}$$

where ε is a small parameter.

The integration in (4.10) and (4.11) is carried over the invariant measure on the group Ω. The functionals Δ_V, Δ_W are obviously invariant under the transformation (4.12).

Using the same arguments as in Chapter 3, we see that on the surface $\nabla_\mu q_\mu = 0$ the functional Δ_V is equal to

$$\Delta_V(\mathscr{A}_\mu, q_\mu)\big|_{\nabla_\mu q_\mu = 0} = \det \varepsilon^{-1} \nabla^2_\mu = = \det[\{\Box - g\partial_\mu[\mathscr{A}_\mu, \ ] - g[\mathscr{A}_\mu, \partial_\mu] + g^2[\mathscr{A}_\mu, [\mathscr{A}_\mu, \ ]]\}] \cdot \varepsilon^{-1}, \tag{4.13}$$

and on the surface $\partial_\mu q_\mu = W$ the functional Δ_W equals

$$\Delta_W(\mathscr{A}_\mu, q_\mu)|_{\partial_\mu q_\mu = W} = \det \varepsilon^{-1}\{\Box - g\partial_\mu[\mathscr{A}_\mu, \ ]\} = \det \varepsilon^{-1} M. \tag{4.14}$$

The constant factors $\det \varepsilon^{-1}$ lead only to the redefinition of the normalizing constant N, and therefore we shall henceforth omit them.

Taking into account (4.10) and (4.14), we rewrite the functional Z_0 as

$$Z_0(A) = N^{-1}\int \exp\left\{ i \int \left[\frac{1}{2} \frac{\delta^2 S_\Lambda}{\delta A_\mu^a(x)\,\delta A_\nu^b(y)} q_\mu^a(x) q_\nu^b(y) + \right.\right.$$
$$\left.\left. + \frac{1}{2\alpha}\{f(\Box) W(x)\}^2 \delta(x-y)\right] dx\, dy \right\} \times$$
$$\times \prod_x \delta(\partial_\mu q_\mu - W)\, \Delta_W(\mathscr{A}_\mu, q_\mu)\, \delta(\nabla_\mu q_\mu^\omega) \times$$
$$\times \Delta_V(\mathscr{A}_\mu, q_\mu)\, d\omega\, dW\, dq_\mu. \tag{4.15}$$

Passing to new variables

$$q_\mu \to q_\mu^{\omega^{-1}}, \quad \omega^{-1} \to \omega \tag{4.16}$$

we have

$$Z_0(A) = N^{-1}\int \exp\left\{ i \int \left[\frac{1}{2} \frac{\delta^2 S_\Lambda}{\delta A_\mu^a(x)\,\delta A_\nu^b(y)} [q_\mu^\omega(x)]^a [q_\nu^\omega(y)]^b + \right.\right.$$
$$\left.\left. + \frac{1}{2\alpha}\{f(\Box) W(x)\}^2 \delta(x-y)\right] dx\, dy \right\} \prod_x \delta(\partial_\mu q_\mu^\omega - W) \times$$
$$\times \Delta_W \delta(\nabla_\mu q_\mu) \det \nabla_\mu^2\, d\omega\, dW\, dq_\mu. \tag{4.17}$$

The δ-function $\delta(\partial_\mu q_\mu^\omega - W)$ removes the integration over ω. The appearing Jacobian cancels out with Δ_W, and ω is expressed in terms of q_μ, W by the equation

$$\partial_\mu(\partial_\mu u - g[\mathscr{A}_\mu, u] + O(u^2)) = \varepsilon(W - \partial_\mu q_\mu). \tag{4.18}$$

The solution of this equation has the form

$$u = \varepsilon M^{-1}(W - \partial_\mu q_\mu) + O(\varepsilon^2). \tag{4.19}$$

Substituting this solution into the formula (4.17), we obtain

$$Z_0(\mathscr{A}) = N^{-1} \int \exp\left\{ i \int \left[\frac{1}{2} \frac{\delta^2 S_\Lambda}{\delta A_\mu^a(x)\, \delta A_\nu^b(y)} \times \right.\right.$$
$$\times [q_\mu + \nabla_\mu M^{-1}(W - \partial_\rho q_\rho)]^a [q_\nu + \nabla_\nu M^{-1}(W - \partial_\rho q_\rho)]^b +$$
$$\left.\left. + \frac{1}{2\alpha}\{f(\Box)\, W(x)\}^2 \delta(x-y)\right] dx\, dy \right\} \times$$
$$\times \det \nabla_\mu^2 \prod_x \delta(\nabla_\mu q_\mu)\, dW\, dq_\mu + O(\varepsilon). \qquad (4.20)$$

Since the functional $Z_0(\mathscr{A})$ in reality is independent of ε, we can set ε in the formula (4.20) equal to 0, as a result of which the last term disappears.

The expression (4.20) differs from (4.8) only in that q_μ in the exponent of the exponential is replaced by

$$q_\mu + \nabla_\mu M^{-1}(W - \partial_\rho q_\rho). \qquad (4.21)$$

Let us show that the additional diagrams appearing as a result of this correspond to convergent integrals. For this we shall take advantage of the relation

$$\int \frac{\delta^2 S_\Lambda}{\delta A_\mu^a(x)\, \delta A_\nu^b(y)} [\nabla_\mu \varphi(x)]^a\, dx = g t^{abc} \frac{\delta S_\Lambda}{\delta A_\nu^q(y)} \varphi^c(y), \quad (4.22)$$

following from the gauge invariance of the action S_Λ. This relation allows us to rewrite the exponent in (4.20) as

$$\frac{1}{2} \int \frac{\delta^2 S_\Lambda}{\delta A_\mu^a(x)\, \delta A_\nu^b(y)} q_\mu^a(x)\, q_\nu^b(y)\, dx\, dy + g t^{abc} \int \frac{\delta S_\Lambda}{\delta A_\nu^a(x)} \times$$
$$\times \left[q_\nu + \frac{1}{2} \nabla_\nu M^{-1}(W - \partial_\rho q_\rho) \right]^b [M^{-1}(W - \partial_\rho q_\rho)]^c\, dx. \qquad (4.23)$$

It is easy to see that the second term in (4.23) generates only convergent diagrams. Indeed, the propagator of q_μ at large q behaves like q^{-6}, and the propagator of W decreases no less rapidly. Therefore the insertion of the vertices generated by the second term in (4.23) in any diagram makes it finite. In particular, the contribution of this term on the mass shell is, in general, equal to zero.

Thus, we have shown that the divergent part of the functional (4.7), which we shall denote by $Z_0'(\mathcal{A})$, can be transformed into the form (4.8). Now let us verify the gauge invariance of this expression.

For this, note that in integrating over q_μ we can write down $Z_0'(A)$ in the form

$$Z_0'(A) = \det Q_0^{-1/2} \det \nabla_\mu^2, \qquad (4.24)$$

where

$$\det Q_0^{-1/2} = \int \exp\left\{ i \int \left[\frac{1}{2} \frac{\delta^2 S_\Lambda}{\delta A_\mu^a(x)\, \delta A_\nu^b(y)} q_\mu^a(x)\, q_\nu^b(y) \right] dx\, dy \right\} \times$$
$$\times \prod_x \delta(\nabla_\mu q_\mu)\, dq_\mu. \qquad (4.25)$$

The functional $\det \nabla_\mu^2$ is manifestly gauge-invariant. Under gauge transformations of the fields $\mathcal{A}_\mu$ the derivatives $\delta^2 S_\Lambda / \delta A_\mu^a(x)\, \delta A_\nu^b(y)$ transforms contragrediently. From the invariance of S_Λ it follows that

$$\frac{\delta^2 S_\Lambda}{\delta A_\mu^a(x)\, \delta A_\nu^b(y)} = \frac{\delta^2 S_\Lambda}{\delta A_\mu'^c(x)\, \delta A_\nu'^d(y)} \times$$
$$\times \left(\delta^{ac}\delta^{bd} - g t^{caf} u^f(x)\, \delta^{bd} - g t^{b\,df} u^f(y)\, \delta^{ac} + \ldots\right). \qquad (4.26)$$

Therefore, if, together with the gauge transformation $\mathcal{A}_\mu \rightarrow \mathcal{A}_\mu^\omega$, the change of integration variables

$$q_\mu \rightarrow \omega q_\mu \omega^{-1}, \qquad (4.27)$$

is performed, then the integral (4.25) will remain unchanged. Thus, finally,

$$Z_0'(\mathcal{A}^\omega) = Z_0'(\mathcal{A}). \qquad (4.28)$$

The regularization consists in the substitution

$$Z_0'(A) \rightarrow \det Q_0^{-1/2} \prod_j \det Q_j^{-\frac{c_j}{2}} \det B_0 \det B_j^{c_j}, \qquad (4.29)$$

where

$$\det B_j = \det \{\nabla_\mu^2 - \mu_j^2\} =$$
$$= \int \exp\left\{-\frac{i}{2}\operatorname{tr}\int [\bar{b}\nabla_\mu^2 b - \mu_j^2 \bar{b} b]\, dx\right\} \prod_x d\bar{b}\, db, \qquad (4.30)$$

$$\det Q_j^{-1/2} = \int \exp\left\{ i \int \left[\frac{1}{2}\int \frac{\delta^2 S_\Lambda}{\delta A_\mu^a(x)\, \delta A_\nu^b(y)} q_\mu^a(x)\, q_\nu^b(y) - \right.\right.$$
$$\left.\left. - \frac{\mu_j^2}{2} q_\mu^2(x)\, \delta(x-y)\right] dx\, dy \right\} \prod_x \delta(\nabla_\mu q_\mu)\, dq_\mu. \qquad (4.31)$$

The variables b and $\bar{b}$ are assumed to anticommute, and the coefficients c_j satisfy the Pauli–Villars conditions

$$\sum c_j + 1 = 0; \quad \sum c_j \mu_j^2 = 0, \qquad (4.32)$$
$$B_0 = B_j \quad \text{at} \quad \mu_j = 0, \quad Q_0 = Q_j \quad \text{at} \quad \mu_j = 0.$$

The formula (4.29) describes the usual Pauli–Villars regularization, which was discussed above in detail for spinor fields as an example. From each closed cycle along which a vector (q_μ) or scalar (b) particle propagates, analogous cycles, in which the internal lines have masses μ_j, are subtracted. Due to the conditions (4.32), the leading terms in the asymptotic expressions of the integrands cancel out and the integrals become convergent. Since $Z_0(\mathscr{A})$ differs from $Z_0'(\mathscr{A})$ only by finite terms, a similar procedure obviously regularizes $Z_0(\mathscr{A})$ also.

The factors det Q_j and det B_j are gauge-invariant. The invariance of det B_j follows directly from the integral representation (4.30). The effective action, occurring in the exponent, describes the gauge-invariant interaction of scalar fields $\bar{b}$, b with the Yang–Mills field. Hence

$$\det B_j(\mathscr{A}^\omega) = \det B_j(\mathscr{A}). \qquad (4.33)$$

The invariance of the functional det Q_j is demonstrated in the same way as the invariance of the functional det Q_0. Thus, the regularization (4.29) is indeed gauge-invariant.

We can now write down the explicit expression for the completely regularized generating functional for the Green functions. It has the form

$$Z_\Lambda(J_\mu) = N^{-1} \int \exp\Big\{ i \int \operatorname{tr}\Big[\frac{1}{8} F_{\mu\nu}F_{\mu\nu} + \frac{1}{8\Lambda^4} \nabla^2 \mathscr{F}_{\mu\nu} \nabla^2 \mathscr{F}_{\mu\nu} -$$

$$- \frac{1}{4\alpha} [f(\Box)\, \partial_\mu \mathscr{A}_\mu]^2 - \frac{1}{2} \mathscr{J}_\mu \mathscr{A}_\mu\Big]\, dx \Big\} \times$$

$$\times \det M \prod_j \det Q_j^{-\frac{c_j}{2}} \det B_j^{c_j} \prod_x d\mathscr{A}_\mu. \tag{4.34}$$

By means of the integral representations (4.30), (4.31) and (III.3.51) this expression can be represented by the path integral of exp{$i \times$ local action} over the Yang–Mills fields $\mathscr{A}_\mu$ and the auxiliary fields $\bar{c}$, c, q_μ, $\bar{b}$, b.

One could also pass in the formula (4.34) to the generalized covariant gauge, by introducing, for example, the term

$$\frac{1}{2\alpha} (\nabla^2 q_\mu^a)^2. \tag{4.35}$$

for fixing the gauge. In addition, the determinants det B_0 and det B_j should be correspondingly modified. We shall not deal with this here.

For finite Λ, μ_j all diagrams generated by Z_Λ converge. At the same time Z_Λ has the same transformation properties as the non-regularized functional formally has. In particular, as we shall see below, the generalized Ward identities are valid for it.

4.5 DIMENSIONAL REGULARIZATION

The divergence index ω depends significantly on the dimensionality of the space-time. For a space of dimension n the product of independent differentials gives to the diagram index a contribution equal to

$$n(L - m + 1), \tag{5.1}$$

where L is the number of internal lines and m is the number of vertices. Therefore, the diagrams to which divergent integrals correspond in a four-dimensional space may turn out to be convergent in a space of smaller dimensionality. On the other hand, going from the four-dimensional space to an n-dimensional space does not influence the symmetry properties. The gauge transformations are generalized in a natural way to a space of any positive dimension. One can go further and define the Feynman diagrams for spaces of noninteger and even complex dimensions. In this case, of course, one cannot speak about any symmetry of the Lagrangian, since the concept itself loses its sense for a noninteger n. Nevertheless, we can investigate the Green functions in a space of arbitrary dimension. As we shall show below, gauge invariance in terms of the Green functions is equivalent to the existence of relations between these functions, known as the generalized Ward identities. These relations make sense in a space of any dimension, and in the region of n where all the integrals converge, they can be proved rigorously. The Green functions, considered as functions of the space dimensionality n, have pole singularities at $n = 4$. Subtracting these singularities, we can continue the Green functions analytically to the point $n = 4$. The functions obtained in this way will satisfy the generalized Ward identities.

As the simplest possible example consider the integral

$$I = \int d^n k \frac{1}{(k^2 - m^2 + i0)\,[(p-k)^2 - m^2 + i0]}. \tag{5.2}$$

For integer $n < 4$ this integral converges. Using the formula

$$\frac{1}{ab} = \int_0^1 \frac{dx}{[ax + b(1-x)]^2}, \tag{5.3}$$

we can rewrite it in the form

$$I = \int_0^1 dx \int d^n k \frac{1}{[k^2 + p^2 x(1-x) - m^2 + i0]^2}. \tag{5.4}$$

Rotating the contour 90° and changing the variables $k_0 \to ik_0$, we come to the integral over the n-dimensional Euclidean space

$$I = i \int_0^1 dx \int d^n k \frac{1}{[k^2 + m^2 - p^2 x(1-x)]^2}. \tag{5.5}$$

The integral over k is readily calculated by using the known formula

$$\int \frac{d^n k}{(k^2 + c)^\alpha} = \frac{\pi^{n/2} c^{\frac{n}{2} - \alpha}\, \Gamma\left(\alpha - \frac{n}{2}\right)}{\Gamma(\alpha)}, \tag{5.6}$$

where $\Gamma(\alpha)$ is the Euler gamma function. The right-hand side of the equality (5.6) can be analytically continued to complex values of n. We shall treat the formula (5.6) as the definition of the integral on the left-hand side for an arbitrary space dimension. Thus, the integral (5.2) is equal to

$$I = i\pi^{\frac{n}{2}}\Gamma\left(2-\frac{n}{2}\right)\int_0^1 dx\,[m^2 - p^2x(1-x)]^{\frac{n}{2}-2}. \tag{5.7}$$

At $n = 4$ the Γ-function has a pole and I tends to infinity. This corresponds to the divergence of the original integral over the four-dimensional space. As usual, the divergence is removed by means of a subtraction procedure. Expanding I in the Laurent series around the point $n/2 = 2$, we have

$$I = \frac{i\pi^2}{\frac{n}{2}-2} - \gamma + i\pi^2\int_0^1 dx \ln[m^2 - p^2(1-x)x] + O\left(\frac{n}{2}-2\right). \tag{5.8}$$

where γ is a finite constant. Performing subtraction at the point $\lambda^2 = p^2$, we obtain the final result in the form

$$\overline{I}(p^2, \lambda^2) = i\pi^2\int_0^1 dx \ln\frac{m^2 - p^2x(1-x)}{m^2 - \lambda^2x(1-x)}. \tag{5.9}$$

For calculations of arbitrary Feynman diagrams it is necessary to formulate also rules for treating tensor entities in an n-dimensional space. By definition,

$$g_{\mu\nu}p_\nu = p_\mu; \quad p_\mu p_\mu = p^2; \quad g_{\mu\nu}g_{\nu\alpha} = \delta_{\mu\alpha}; \quad g_{\mu\nu}g_{\mu\nu} = n. \tag{5.10}$$

In a similar manner, for theories including fermions, objects are introduced which possess the algebraic properties of the γ-matrices:

$$\gamma_\mu\gamma_\nu + \gamma_\nu\gamma_\mu = 2g_{\mu\nu}I, \tag{5.11}$$

where I is the unit matrix, and

$$\mathrm{tr}\,\{\gamma_\mu\gamma_\nu\} = 2^{\frac{n}{2}}g_{\mu\nu}, \tag{5.12}$$

$$\gamma_\mu\hat{p}\gamma_\mu = 2\left(1-\frac{n}{2}\right)\hat{p}, \quad \gamma_\mu\hat{p}\hat{q}\gamma_\mu = 4pq + (n-4)\,\hat{p}\hat{q}. \tag{5.13}$$

Note, however, that the usual definition of the matrix γ_5,

$$\gamma_5 = \frac{1}{4!} \varepsilon_{\alpha\beta\mu\nu} \gamma_\alpha \gamma_\beta \gamma_\mu \gamma_\nu \tag{5.14}$$

is not applicable in a space of arbitrary dimension, since the completely antisymmetric tensor $\varepsilon_{\alpha\beta\mu\nu}$ is defined only in the four-dimensional space. Because of this, the theories to which the matrix γ_5 pertains need special consideration, and in general the dimensional regularization is not applicable to them.

The recipe for the dimensional regularization of an arbitrary Feynman diagram consists in the following. In the integral corresponding to the diagram

$$F = \prod_{j=1}^{I} \int d^n k_j \left(k_{l_1}\right)_\lambda \left(k_{l_2}\right)_\mu \cdots \left(k_{l_n}\right)_\nu \prod_{i=1}^{L} (q_i^2 - m_i^2 + i0), \tag{5.15}$$

where L is the number of internal lines in the diagram, I is the number of independent cycles, and the momenta q_i represent the algebraic sums of the integration variables k_i and the external momenta p_i, it is necessary to use a parametrization of the Green functions which allows us to perform the integration over k_i explicitly. For this it is possible to use either the Feynman parametrization (5.3) or the so-called α-representation

$$(p^2 - m^2 + i0)^{-1} = (i)^{-1} \int_0^\infty d\alpha \exp\{i\alpha (p^2 - m^2 + i0)\}. \tag{5.16}$$

After passing to the α-representation, the integrals over k become Gaussian and are calculated by formulas of the type

$$\int \frac{d^n k}{(2\pi)^n} \exp(-xk^2 + 2ka) = \left(\frac{\pi}{x}\right)^{\frac{n}{2}} (2\pi)^{-n} \exp\left\{\frac{a^2}{x}\right\}; \tag{5.17}$$

for noninteger n the formula (5.17) is considered to be the definition of the integral over the n-dimensional space.

The integral defining the function F converges in a finite region of the complex variables n. At $n = 4$ this function has poles. (In practice these poles appear, for example, as singularities of the Euler Γ-functions, which result from integration over the parameters α.)

By singling out the tensor structures by means of the combinatorial formulas (5.10), the functions F can be represented as a sum of scalar functions F_i. If the corresponding diagram has no divergent subgraphs, then the Laurent expansion of the functions F_i around the point $n = 4$ has the form

$$F_i(p) = \frac{A(p_i^2, p_i p_j)}{n-4} + B(p_i^2, p_i p_j) + O(n-4), \qquad (5.18)$$

where $A(p_i^2, p_i p_j)$ is a polynomial of order equal to the diagram index.

Subtracting from the function $F_j(p)$ several leading terms of its expansion in the Taylor series in the momenta p_j, we obtain a function which can be analytically continued to the point $n = 4$.

For diagrams containing divergent subgraphs the subtractions are performed successively. Counterterms are introduced, which remove the divergence from the subgraphs, and then a subtraction is performed for the diagram as a whole. An important property of the dimensional regularization is the possibility of shifting the integration variables within regularized integrals. It is just this property, together with the tensor algebra (5.10), which allows one to prove the generalized Ward identities within the framework of dimensional regularization.

In conclusion we shall illustrate the dimensional-regularization method by a simple example—the calculation of second-order corrections to the Green function of the Yang–Mills field. This correction is described by the diagrams in Figure 4.

To the diagram (a) in the diagonal α-gauge there corresponds the regularized integral

$$\begin{aligned}
\Pi^{ab}_{\mu\nu}(p)_a = & -\frac{g_1^2}{2} \int \frac{d^n k}{(2\pi)^n} \varepsilon^{aa_1a_2} \varepsilon^{bb_1b_2} \times \\
& \times [(p+k)_{\mu_2} g_{\mu\mu_1} + (p-2k)_{\mu} g_{\mu_1\mu_2} + (k-2p)_{\mu_1} g_{\mu\mu_2}] \times \\
& \times [(k+p)_{\nu_2} g_{\nu\nu_1} + (k-2p)_{\nu_1} g_{\nu\nu_2} + (p-2k)_{\nu} g_{\nu_1\nu_2}] \times \\
& \times \frac{(-i\delta^{a_1b_1})}{k^2+i0} g_{\mu_1\nu_1} \frac{(-i\delta^{a_2b_2})}{(p-k)^2+i0} g_{\mu_2\nu_2} = \\
= & \, g_1^2 \delta^{ab} \int \frac{d^n k}{(2\pi)^n} \{g_{\mu\nu}[(k+p)^2 + (k-2p)^2] + (n-6) p_\mu p_\nu + \\
& + (4n-6) k_\mu k_\nu + (3-2n)(p_\nu k_\mu + p_\mu k_\nu)\} \times \\
& \times \{(k^2+i0)[(p-k)^2+i0]\}^{-1},
\end{aligned} \qquad (5.19)$$

where, in order to conserve the correct dimensionality of $\Pi_{\mu\nu}$, a dimensional coupling constant $g_1^2 = g^2\mu^{4-n}$ has been introduced.

Using the formula

$$\frac{1}{k^2 (p-k)^2} = \int_0^1 dz \frac{1}{[k^2(1-z)+(p-k)^2 z]^2} \tag{5.20}$$

and passing to new variables

$$k \to k + pz, \tag{5.21}$$

we write this integral in the form

$$\Pi_{\mu\nu}^{ab}(p)_a = g_1^2 \delta^{ab} \int_0^1 dz \frac{d^n k}{(2\pi)^n} [g_{\mu\nu}(5 - 2z + 2z^2) p^2 + 2g_{\mu\nu}k^2 +$$
$$+ (4n-6) k_\mu k_\nu - (4n-6) z(1-z) p_\mu p_\nu +$$
$$+ (n-6) p_\mu p_\nu][k^2 + p^2 z(1-z) + i0]^{-1}. \tag{5.22}$$

In this formula the terms odd in k are dropped, since their contribution vanishes for symmetry reasons. Passing to the Euclidean metric, it is possible to integrate over k by the formula

$$\int \frac{(k^2)^m d^n k}{[k^2 + p^2 z(1-z)]^l} = \frac{i\pi^{n/2}}{\Gamma\left(\frac{n}{2}\right)} \int_0^\infty dk^2 \frac{(-k^2)^m (k^2)^{\frac{n}{2}-1}}{[-k^2 + p^2 z(1-z)]^l} =$$
$$= \frac{i\pi^{\frac{n}{2}}}{\Gamma\left(\frac{n}{2}\right)} (-1)^{m+l} [-p^2 z(1-z)]^{m+\frac{n}{2}-l} \times$$
$$\times \frac{\Gamma\left(m+\frac{n}{2}\right) \Gamma\left(l-\frac{n}{2}-m\right)}{\Gamma(l)}. \tag{5.23}$$

For a noninteger or complex n this formula represents the definition of the integral in the left-hand side of (5.23). Integrating, we obtain

$$\Pi^{ab}_{\mu\nu}(p)_a = \frac{ig_1^2\delta^{ab}}{(4\pi)^{\frac{n}{2}}}\int_0^1 dz\Big\{[g_{\mu\nu}(5-2z+2z^2)p^2 - $$

$$-(4n-6)p_\mu p_\nu z(1-z)+(n-6)p_\mu p_\nu]\times$$

$$\times[-p^2z(1-z)]^{\frac{n}{2}-2}\,\Gamma\Big(2-\frac{n}{2}\Big)-$$

$$-3(n-1)g_{\mu\nu}[-p^2z(1-z)]^{\frac{n}{2}-1}\,\Gamma\Big(1-\frac{n}{2}\Big). \quad (5.24)$$

Integration over z is performed using the formula

$$\int_0^1 dz\, z^{m-n-1}(1-z)^{m-k-1} = \frac{\Gamma(m-n)\,\Gamma(m-k)}{\Gamma(2m-n-k)}. \quad (5.25)$$

As a result we obtain

$$\Pi^{ab}_{\mu\nu}(p)_a = \frac{ig_1^2}{(4\pi)^{n/2}}\,\delta^{ab}\Big\{g_{\mu\nu}p^2\Big[5\,\frac{\Gamma^2\left(\frac{n}{2}-1\right)}{\Gamma(n-2)}-$$

$$-2\,\frac{\Gamma\left(\frac{n}{2}\right)\Gamma\left(\frac{n}{2}-1\right)}{\Gamma(n-1)}+2\,\frac{\Gamma\left(\frac{n}{2}+1\right)\Gamma\left(\frac{n}{2}-1\right)}{\Gamma(n)}+$$

$$+\frac{6(n-1)}{2-n}\,\frac{\Gamma^2\left(\frac{n}{2}\right)}{\Gamma(n)}\Big]-p_\mu p_\nu\Big[(4n-6)\frac{\Gamma^2\left(\frac{n}{2}\right)}{\Gamma(n)}-$$

$$-(n-6)\,\frac{\Gamma^2\left(\frac{n}{2}-1\right)}{\Gamma(n-2)}\Big]\Big\}\,\Gamma\Big(2-\frac{n}{2}\Big)(-p^2)^{\frac{n}{2}-2}. \quad (5.26)$$

In deriving this formula we have used the relation

$$\Gamma(1-\omega) = \frac{1}{(1-\omega)}\,\Gamma(2-\omega). \quad (5.27)$$

As is seen, in the limit $n \to 4$

$$\Pi_{\mu\nu}(p)_a \to \infty,$$

since the function $\Gamma(2 - n/2)$ has a pole at this point. Expanding $\Pi_{\mu\nu}(p)_a$ around $n = 4$ in the Laurent series and taking into account that

$$\left(\frac{\mu^2}{-p^2}\right)^\varepsilon = 1 + \varepsilon \ln\left(\frac{\mu^2}{-p^2}\right) + O(\varepsilon^2), \tag{5.28}$$

we obtain the final expression for $\Pi_{\mu\nu}(p)_a$:

$$\begin{aligned}
&\Pi^{ab}_{\mu\nu}(p)_a = \\
&= \frac{ig^2\delta^{ab}}{16\pi^2}\left\{(g_{\mu\nu}p^2 - p_\mu p_\nu)\left(\frac{19}{6}\varepsilon^{-1} + c_a\right) - \frac{1}{2}p_\mu p_\nu(d + \varepsilon^{-1}) + \right. \\
&\quad \left. + (g_{\mu\nu}p^2 - p_\mu p_\nu)\frac{19}{6}\ln\frac{\mu^2}{-p^2} - \frac{1}{2}p_\mu p_\nu \ln\frac{\mu^2}{-p^2}\right\}; \\
&\qquad\qquad \varepsilon = \frac{4-n}{2},
\end{aligned} \tag{5.29}$$

where c_a and d are finite constants.

The integral

$$\Pi^{ab}_{\mu\nu}(p)_b = -2g_1^2 \int \frac{d^n k}{(2\pi)^n}\frac{k_\mu(k-p)_\nu}{k^2(p-k)^2}\delta^{ab}. \tag{5.30}$$

corresponds to the diagram (b). Calculations completely analogous to the preceding ones give

$$\begin{aligned}
\Pi^{ab}_{\mu\nu}(p)_b = \frac{ig^2}{16\pi^2}\delta^{ab}&\left\{(g_{\mu\nu}p^2 - p_\mu p_\nu)\left(\frac{1}{6}\varepsilon^{-1} + c_b\right) + \right. \\
&+ \frac{1}{2}p_\mu p_\nu(\varepsilon^{-1} + d) + (g_{\mu\nu}p^2 - p_\mu p_\nu)\frac{1}{6}\ln\frac{\mu^2}{-p^2} + \\
&\left. + \frac{1}{2}p_\mu p_\nu \ln\frac{\mu^2}{-p^2}\right\}.
\end{aligned} \tag{5.31}$$

And, finally, the diagram (c) gives a zero contribution. The contribution of this diagram is proportional to the integral

$$I = \int \frac{d^4 k}{k^2}. \tag{5.32}$$

In the method of dimensional regularization the formula

$$I = \int \frac{d^n k\,(k^2)^{\alpha-1}}{(2\pi)^n} = 0; \qquad \alpha = 0, 1, \ldots, m. \tag{5.33}$$

is valid. Thus, the total second-order correction to the Green function of the Yang–Mills field has the form

$$\Pi^{ab}_{\mu\nu}(p) = \Pi^{ab}_{\mu\nu}(p)_a + \Pi^{ab}_{\mu\nu}(p)_b =$$
$$= \frac{ig^2\delta^{ab}}{16\pi^2}(g_{\mu\nu}p^2 - p_\mu p_\nu)\left\{\frac{10}{3}\varepsilon^{-1} + c + \frac{10}{3}\ln\frac{\mu^2}{-p^2}\right\}. \quad (5.34)$$

The divergence as $\varepsilon \to 0$ is removed, as usual, by a subtraction procedure. The corresponding counterterm is equal to

$$(z_2 - 1) = \frac{5g^2}{24\pi^2}\varepsilon^{-1} + c, \quad (5.35)$$

in agreement with the formula (1.33) from Section 4.1. As is seen, the expression (5.34) is automatically transverse, and for the removal of divergences there is no need of gauge–noninvariant counterterms, such as the counterterm for the mass renormalization of the Yang–Mills field.

4.6 GENERALIZED WARD IDENTITIES

The renormalization procedure is usually formulated in terms of the Green functions. Unlike the S-matrix, the Green functions are not gauge-invariant objects, and their values depend on the specific gauge condition chosen. The relativity principle is equivalent to the existence of relations between the Green functions, which we shall, by analogy with electrodynamics, call the generalized Ward identities. These relations provide for the physical equivalence of various gauges and play a key role in the proof of the gauge invariance and the unitarity of the renormalized S-matrix. From them it follows, in particular, that the counterterms needed for the removal of the intermediate regularization form a gauge-invariant structure.

We shall start with the derivation of the generalized Ward identities for regularized nonrenormalized Green functions. In all further reasoning only the gauge invariance of the regularized action will be used. We shall not, therefore, write out its explicit expression, having in mind that we can always use for this purpose, for example, the formula (4.34).

As the initial representation of the generating functional for the

Green functions we shall choose

$$Z = N^{-1} \int \exp\left\{ i \left[S_\Lambda - \right.\right.$$
$$\left.\left. - \frac{1}{2} \operatorname{tr} \int \left[\frac{1}{2\alpha} (f(\Box) W(x))^2 + \mathcal{J}_\mu \mathcal{A}_\mu \right] dx \right] \right\} \times$$
$$\times \Delta(\mathcal{A}) \prod_x \delta(\partial_\mu \mathcal{A}_\mu - W) \, d\mathcal{A} \, dW. \quad (6.1)$$

Here S_Λ is the gauge-invariant action functional, which contains all the regularizing factors. For obtaining the generalized Ward identities we shall use the same method as was used for the proof of the gauge invariance of the S-matrix.

Let us introduce the gauge-invariant function $\tilde{\Delta}(\mathcal{A})$ defined by the condition

$$\tilde{\Delta}(\mathcal{A}) \int \delta \left[\partial_\mu \mathcal{A}_\mu^\omega - W(x) - \chi(x) \right] d\omega = 1, \quad (6.2)$$

where $\chi(x)$ is an arbitrary matrix function. Allowing for (6.2), we rewrite $Z(J)$ as

$$Z(J) = N^{-1} \int \exp\left\{ i \left[S_\Lambda - \frac{1}{2} \operatorname{tr} \int \left[\mathcal{J}_\mu \mathcal{A}_\mu + \frac{1}{2\alpha} (f(\Box) w)^2 \right] dx \right] \right\} \times$$
$$\times \Delta(\mathcal{A}) \tilde{\Delta}(\mathcal{A}) \prod_x \delta(\partial_\mu \mathcal{A}_\mu - W) \times$$
$$\times \delta\left(\partial_\mu \mathcal{A}_\mu^\omega - W - \chi \right) d\mathcal{A} \, dW \, d\omega. \quad (6.3)$$

We pass to new variables

$$\begin{aligned} \mathcal{A}_\mu &\to \mathcal{A}_\mu^\omega, \\ \omega &\to \omega^{-1}. \end{aligned} \quad (6.4)$$

The integrals over ω and W are removed by δ-functions, and the Jacobian appearing cancels with $\Delta(\mathcal{A})$.

Taking into account that the value of the functional $\tilde{\Delta}(\mathcal{A})$ on the surface,

$$\partial_\mu \mathcal{A}_\mu = W + \chi \quad (6.5)$$

is equal to the value of the functional $\Delta(\mathscr{A})$ on the surface,

$$\partial_\mu \mathscr{A}_\mu = W, \tag{6.6}$$

we obtain

$$Z(J) = N^{-1} \int \exp\left\{ i \left[S_\Lambda - \frac{1}{2} \operatorname{tr} \int \left[J_\mu \mathscr{A}_\mu^\omega + \right.\right.\right.$$
$$\left.\left.\left. + \frac{1}{2\alpha} (f(\square)(\partial_\mu \mathscr{A}_\mu - \chi))^2 \right] dx \right\} \det M \prod_x d\mathscr{A}. \tag{6.7}$$

Here

$$\mathscr{A}_\mu^\omega = \mathscr{A}_\mu + \partial_\mu u - g[\mathscr{A}_\mu, u] + O(u^2), \tag{6.8}$$

and $u(x)$ satisfies the equation

$$\square u - g\partial_\mu [\mathscr{A}_\mu, u] + O(u^2) = W - \partial_\mu \mathscr{A}_\mu = -\chi. \tag{6.9}$$

Representing the solution u by a series in χ, we have

$$u = -M^{-1}\chi + O(\chi^2), \tag{6.10}$$

where M^{-1} is the inverse operator of M. The kernel of this operator, $M^{-1}_{ab}(x, y)$, satisfies the equation

$$\square M_{ab}^{-1}(x, y) - g t^{adc} \partial_\mu \left(A_\mu^d(x) M_{cb}^{-1}(x, y) \right) =$$
$$= \delta^{ab} \delta(x - y) \tag{6.11}$$

and obviously coincides with the connected part of the Green function of the fictitious particles in an external classical field $\mathscr{A}_\mu(x)$:

$\mathscr{A}_\mu(x)$:

$$M_{ab}^{-1}(x, y) = \delta^{ab} D^0(x - y) +$$
$$+ g t^{adb} \int D^0(x - z) \partial_\mu \left[A_\mu^d(z) D^0(z - y) \right] dz + \ldots \tag{6.12}$$

Since the original functional (6.1) does not depend on χ, its derivative with respect to χ is equal to zero

$$\left. \frac{dZ}{d\chi} \right|_{\chi=0} = 0. \tag{6.13}$$

Substituting into this formula the expression (6.7), obtained by the identity transformation of the functional (6.1), and performing the differentiation explicitly, we get

$$\int \exp\left\{i\left[S_\Lambda + \int\left[J^a_\mu A^a_\mu + \frac{1}{2\alpha}(f(\Box)\,\partial_\mu A_\mu)^2\right]dx\right\}\det M \times$$
$$\times\left\{\frac{1}{\alpha}f^2(\Box)\,\partial_\mu A^a_\mu(y) + \right.$$
$$\left. + \int J^b_\mu(z)\left(\nabla^z_\mu M^{-1}\right)^{ba}(z,\, y,\, \mathscr{A})\,dz\right\}\prod_x d\mathscr{A} = 0. \qquad (6.14)$$

This equality is nothing but the system of generalized Ward identities for the Yang–Mills theory. It can be written also in terms of variational derivatives

$$\left\{\frac{1}{\alpha}f^2(\Box)\,\partial_\mu\left\{\frac{1}{i}\frac{\delta}{\delta J^a_\mu(x)}\right\} + \right.$$
$$\left. + \int J^b_\mu(y)\left[\nabla^y_\mu\left(\frac{1}{i}\frac{\delta}{\delta J_\mu(y)}\right)M^{-1}_{yx}\left(\frac{1}{i}\frac{\delta}{\delta J}\right)\right]^{ba}dy\right\}Z = 0, \qquad (6.15)$$

where the operators $(M^{-1})^{ba}\left(\frac{1}{i}\frac{\delta}{\delta J}\right)$ and $\nabla_\mu\left(\frac{1}{i}\frac{\delta}{\delta J}\right)$ are derived from $\nabla_\mu(A)$, $(M^{-1})^{ba}(x, y, \mathscr{A})$ by the obvious substitution

$$A^a_\mu \to \frac{1}{i}\frac{\delta}{\delta J^a_\mu}. \qquad (6.16)$$

Applying the operator M^{-1} to $Z(J)$, we obtain the total Green function of the fictitious particles in the presence of the classical source J:

$$(M^{-1})^{ab}_{xy}\left[\frac{1}{i}\frac{\delta}{\delta J}\right]Z(J) = G^{ab}(x,\, y,\, J) =$$
$$= -\frac{\delta^2}{\delta\bar\eta^a(x)\,\delta\eta^b(y)}N^{-1}\int\exp\left\{-i\,\frac{\mathrm{tr}}{2}\left[\bar{c}\partial_\mu\nabla_\mu c + \mathscr{L}_{YM} + \right.\right.$$
$$\left.\left. + \bar{c}\eta + \bar\eta c + \mathscr{J}_\mu\mathscr{A}_\mu\right]dx\right\}\prod_x d\bar{c}\,dc\,d\mathscr{A}\Big|_{\bar\eta=\eta=0}. \qquad (6.17)$$

This function satisfies the equation

$$\partial^x_\mu\nabla^{ab}_\mu\left(\frac{1}{i}\frac{\delta}{\delta J}\right)G^{bc}(x,\, y,\, J) = \delta^{ac}\delta(x-y)\,Z(J). \qquad (6.18)$$

From the generalized Ward identities (6.15) it is easy to obtain relations between various Green functions. For instance, performing variational differentiation of (6.15) with respect to $J_\nu^b(y)$ at $J = 0$ and differentiating the resulting equality with respect to y_ν, we get

$$\frac{-i}{\alpha} f^2(\Box)\,\partial_\mu^x \partial_\nu^y \left[\frac{\delta^2 Z}{\delta J_\mu^a(x)\,\delta J_\nu^b(y)}\right]_{J=0} =$$
$$= -\partial_\nu^y \left\{\nabla_\nu^y \left(\frac{1}{i}\frac{\delta}{\delta J}\right) G_{yx}\right\}_{J=0}^{ba} = -\delta^{ab}\delta(x-y). \qquad (6.19)$$

The variational derivative

$$\frac{1}{i}\frac{\delta^2 Z}{\delta J_\mu^a(x)\,\delta J_\nu^b(y)}\Bigg|_{J=0} \qquad (6.20)$$

is just the two-point Green function of the Yang–Mills field $G_{\mu\nu}^{ab}(x, y)$. The equality (6.19) shows that the longitudinal part of the complete Green function,

$$G_{\mu\nu}^L(x-y) = \partial_\mu \partial_\nu \Box^{-2} \partial_\rho \partial_\sigma G_{\rho\sigma}(x-y), \qquad (6.21)$$

coincides with the free one,

$$G_{\mu\nu}^L = D_{\mu\nu}^L = -\delta^{ab}\frac{\alpha}{(2\pi)^4}\int e^{-ikx}\frac{k_\mu k_\nu}{k^4 f^2(-k^2)}\,dk. \qquad (6.22)$$

Thus, in complete analogy with electrodynamics, radiative corrections to the longitudinal part of the Green function are absent. The corresponding identities for three- and four-point functions look significantly more complicated than in electrodynamics, since they include in a nontrivial manner the Green functions of fictitious particles.

The consequence of the generalized Ward identities is the existence of relations between counterterms, necessary for the removal of divergences from the Green functions. For example, from the identity (6.19) for the two-point Green function it follows that the counterterm responsible for the renormalization of the longitudinal part of the wave function is equal to zero. It can be shown that if the Green functions satisfy the generalized Ward identities, then the counterterms form a gauge-invariant structure. This may be done either by directly analyzing the system (6.15) or by passing to analogous identities for one-particle irreducible Green functions. By this it is proved that renormalization does not violate the gauge invariance of the theory. It is simpler, however, to proceed the other way round—from the very beginning to insert in the Lagrangian gauge-invariant counterterms of the most general type, and then, using the generalized Ward identities, to prove that all the Green functions in such a theory tend to a finite limit when the intermediate regularization is removed. That is exactly what we shall do in the next section.

Let us now obtain the generalized Ward identities for the case when the Yang–Mills field interacts with scalar φ and spinor ψ fields.

The generating functional for the Green functions in this case can be written as

$$Z(J, \zeta, \bar{\eta}, \eta) = N^{-1}\int \exp\Big\{i\Big[S_\Lambda + \int\Big[\frac{1}{2\alpha}(f(\Box)W)^2 + \\ + J_\mu^a A_\mu^a + \zeta^i\varphi^i + \bar{\psi}^k\eta^k + \bar{\eta}^k\psi^k\Big]dx\Big\}\times \\ \times\Delta(\mathcal{A})\prod_x \delta(\partial_\mu\mathcal{A}_\mu - W)\,d\mathcal{A}\,dW\,d\varphi\,d\bar{\psi}\,d\psi. \quad (6.23)$$

The gauge-invariant regularized action includes also terms describing the interaction of spinor and scalar fields.

All the reasoning given above automatically applies to this case. The only difference is that, in addition to the change of variables (6.4), it is necessary to pass to new fields φ^ω, ψ^ω

$$\varphi \to \varphi^\omega, \quad \psi \to \psi^\omega. \quad (6.24)$$

As a result, there appear in the exponent in the transformed functional the additional terms

$$\delta_\chi(\zeta^i\varphi^i + \bar{\psi}^k\eta^k + \bar{\eta}^k\psi^k) = \zeta^i\delta_\chi\varphi^i + \delta_\chi\bar{\psi}^k\eta^k + \bar{\eta}^k\delta\psi^k \quad (6.25)$$

and the generalized Ward identities take the form

$$\int \exp\Big\{i\Big[S_\Lambda + \int\Big[A_\mu^a J_\mu^a + \zeta^i\varphi^i + \bar{\psi}^j\eta^j + \bar{\eta}^j\psi^j + \\ + \frac{1}{2\alpha}(f(\Box)\,\partial_\mu A_\mu^a)^2\Big]dx\Big\}\det M\Big\{\frac{1}{\alpha}f^2(\Box)\,\partial_\mu A_\mu^a(y) + \\ + \int\Big[J_\mu^b(z)\,[\nabla_\mu^z M^{-1}(z, y, \mathcal{A})]^{ba} + \zeta^i(z)\frac{\delta_\chi\varphi^i(z)}{\delta\chi^a(y)}\Big|_{\chi=0} + \\ + \frac{\delta_\chi\bar{\psi}^k(z)}{\delta\chi^a(y)}\Big|_{\chi=0}\eta^k(z) + \bar{\eta}^k(z)\frac{\delta_\chi\psi^k(z)}{\delta\chi^a(y)}\Big|_{\chi=0}\Big]dz\Big\} = 0. \quad (6.26)$$

In this form the generalized Ward identities are valid both in the symmetric theory and in the theory with spontaneous symmetry breaking. The difference is only in the explicit form of the gauge transformation of the scalar fields φ^{ω}. If the fields φ realize a representation of the gauge group Ω with the generators Γ^c,

$$\delta\varphi^a = g(\Gamma^c)^{ab}\varphi^b u^c + O(u^2), \qquad (6.27)$$

then the identity (6.26) can be written in the form (we omit the spinor fields)

$$\frac{1}{\alpha} f^2(\Box)\,\partial^x_\mu\left[\frac{1}{i}\frac{\delta Z}{\delta J^a_\mu(x)}\right] + \\ + \int\Big[J^b_\mu(y)\left\{\nabla_\mu\left[\frac{1}{i}\frac{\delta}{\delta J}\right]G(y, x, J, \zeta)\right\}^{ba} + \\ + g\zeta^b(y)(\Gamma^d)^{bc}\frac{1}{i}\frac{\delta G^{da}(y, x, J, \zeta)}{\delta\zeta^c(y)}\Big]dy = 0. \qquad (6.28)$$

In the case of spontaneously broken symmetry the transformation (6.27) is modified in the following manner:

$$\delta\varphi^a = g(\Gamma^c)^{ab}\,\varphi^b u^c + g(\Gamma^c)^{ab}\,r^b u^c + O(u^2), \qquad (6.29)$$

where r^b is a constant vector, which may be considered without loss of generality to be directed along the axis labeled $\bar{b}$: $r^b = r\delta^{\bar{b}b}$. Correspondingly there appears in the identity (6.28) an additional term

$$rg\int\zeta^b(y)\,(\Gamma^d)^{b\bar{b}}\,G^{da}(y, x, J, \zeta)\,dy. \qquad (6.30)$$

For instance, for the model (1.3.25) in which the scalar fields form the complex SU_2 doublet

$$\varphi(x) = \begin{pmatrix}\varphi_1(x)\\ \varphi_2(x)\end{pmatrix} = \frac{1}{\sqrt{2}}\begin{pmatrix} iB_1(x) + B_2(x)\\ \sqrt{2}\,\mu + \sigma(x) - iB_3(x)\end{pmatrix}, \qquad (6.31)$$

the gauge transformation has the form

$$\delta\sigma = \frac{-g}{2}(B^a u^a),$$
$$\delta B^a = -m_1 u^a - \frac{g}{2}\varepsilon^{abc}B^b u^c - \frac{g}{2}\sigma u^a, \qquad (6.32)$$
$$m_1 = \frac{g\mu}{\sqrt{2}}.$$

The generalized Ward identities appear as follows:

$$\frac{1}{\alpha} f^2(\Box)\,\partial_\mu \left[\frac{1}{i}\frac{\delta Z}{\delta J^a_\mu(x)}\right] + \int \Big\{\Big[J^b_\mu(y)\,\nabla^{bd}_\mu\Big(\frac{1}{i}\frac{\delta}{\delta J}\Big) -$$

$$- \zeta_\sigma(y)\frac{g}{2}\frac{1}{i}\frac{\delta}{\delta\zeta^d_B(y)} - \zeta^b_B(y)\Big(\frac{g}{2}\,\varepsilon^{bcd}\frac{1}{i}\frac{\delta}{\delta\zeta^c_B(y)} +$$

$$+ \delta^{bd}\frac{g}{2}\frac{1}{i}\frac{\delta}{\delta\zeta_\sigma(y)} + m_1\delta^{bd}\Big)\Big]\, G^{da}(y, x, J, \zeta)\Big\}\, dy = 0. \qquad (6.33)$$

To conclude this section we shall show that the generalized Ward identities (6.14) express a certain additional symmetry, having no classical analog, of the effective Lagrangian of the quantized Yang–Mills field. [By the effective Lagrangian we mean the expression in the exponent in the formula (3.3.54) for the S-matrix. This expression contains, besides the classical Yang–Mills Lagrangian, a gauge-fixing term and the Lagrangian of fictitious fields.]

Let us write the generating functional for the Green functions in the form of a path integral of $\exp\{i \times \text{local action}\}$, introducing explicitly the fields of fictitious particles:

$$Z = N^{-1}\int \exp\Big\{ i \int \Big[-\frac{1}{4}F^a_{\mu\nu}F^a_{\mu\nu} + \frac{1}{2\alpha}(\partial_\mu A_\mu)^2 + \bar{c}^a M^{ab} c^b +$$

$$+ \bar{c}^a\eta^a + \bar{\eta}^a c^a\Big] dx \Big\} \prod_x d\mathcal{A}\, d\bar{c}\, dc. \qquad (6.34)$$

In this formula, besides the source for the Yang–Mills field, we have introduced anticommuting sources $\bar{\eta}$, η for the fictitious fields. The functional (6.34) corresponds to a definite choice of the gauge condition [for simplicity we consider the case $f(\Box) = 1$]. Therefore the effective action in the exponent is not gauge-invariant. Nevertheless, there exist transformations which affect simultaneously both the Yang–Mills fields and the fictitious fields $\bar{c}$, c with respect to which the effective Lagrangian is invariant. These transformations have the following form:

$$A^a_\mu(x) \to A^a_\mu(x) + [\nabla_\mu c(x)]^a\,\varepsilon, \qquad (6.35)$$

$$c^a(x) \to c^a(x) - \frac{1}{2} t^{abd} c^b(x)\, c^d(x)\,\varepsilon, \qquad (6.36)$$

$$\bar{c}^a(x) \to \bar{c}^a(x) + \frac{1}{\alpha}\left[\partial_\mu A^a_\mu(x)\right]\varepsilon. \qquad (6.37)$$

Here ε is a parameter independent of x and is an element of the Grassman algebra:

$$\varepsilon^2 = 0; \quad \varepsilon c + c\varepsilon = 0; \quad \varepsilon\bar{c} + \bar{c}\varepsilon = 0; \quad [\varepsilon, \mathscr{A}_\mu] = 0. \quad (6.38)$$

(We recall that the fictitious fields $\bar{c}$, c are also anticommuting variables.) Such transformations, which mix up commuting and anticommuting quantities in a nontrivial manner, have become known as supertransformations.

Let us verify that the effective Lagrangian figuring in the formula (6.34) is invariant under the transformations (6.35) to (6.37). The transformation (6.35) is a special case of the gauge transformation, because it leaves the first term in the exponent of (6.34) invariant. It is not difficult to check that the variation $\delta(\nabla_\mu c)$ is also equal to zero. Indeed,

$$\delta(\nabla_\mu c)^a = -\frac{1}{2} t^{abd} \left[\partial_\mu (c^b c^d) - A_\mu^b t^{def} c^e c^f\right] \varepsilon - \\ - t^{abd} \left(\partial_\mu c^b - t^{bef} A_\mu^e c^f\right) \varepsilon c^d. \quad (6.39)$$

The anticommutativity of the variables c^f, c^d leads to

$$t^{abd} t^{bef} c^f c^d = \frac{1}{2} t^{abd} t^{def} c^e c^f, \quad (6.40)$$

and therefore the right-hand side of (6.39) vanishes. Thus, the total variation of the effective Lagrangian is equal to

$$\delta\mathscr{L}_э = \frac{1}{\alpha} \partial_\mu A_\mu^a \partial_\mu (\nabla_\mu c)^a \varepsilon - \frac{1}{\alpha} \partial_\mu A_\mu^a M^{ab} c^b \varepsilon. \quad (6.41)$$

Remembering the definition of the operator M, we see that this expression is equal to zero.

The transformations (6.35) to (6.37) do not have any clear geometrical meaning, and the invariance of the effective Lagrangian with respect to these transformations is not connected with the conservation of any observable quantity. Nevertheless it leads to a number of useful consequences and, in particular, can be used for the alternative derivation of the generalized Ward identities.

For this purpose, in the integral (6.34) we shall make the change of integration variables (6.35) to (6.37). The Jacobian of this transformation, which can be symbolically written as

$$J = \det \begin{pmatrix} \delta_{\mu\nu}\delta^{ab}, & \nabla_{\mu}^{ab}\varepsilon, & 0 \\ 0, & \delta^{ab} - t^{adb}c^{d}\varepsilon, & 0 \\ \delta^{ab}\dfrac{1}{\alpha}\,\varepsilon\partial_{\mu}, & 0, & \delta^{ab} \end{pmatrix}, \tag{6.42}$$

is obviously equal to unity. Therefore, as a result of the substitution (6.35) to (6.37) only the terms with sources are changed. Writing out their variation explicitly and equating the derivative $dZ/d\varepsilon$ to zero, we obtain the relation

$$0 = \int \exp\left\{ i \int [\mathcal{L}_{\vartheta} + J_{\mu}^{a}A_{\mu}^{a} + \bar{c}^{a}\eta^{a} + \bar{\eta}^{a}c^{a}]\,dx \right\} \times$$
$$\times \left\{ J_{\mu}^{b}(y)\,[\nabla_{\mu}c(y)]^{b} - \frac{1}{\alpha}\partial_{\mu}A_{\mu}^{a}(y)\,\eta^{a}(y) - \right.$$
$$\left. - \frac{1}{2}\bar{\eta}^{a}(y)\,t^{abd}c^{b}(y)\,c^{d}(y) \right\} dy \prod_{x} d\mathcal{A}\,d\bar{c}\,dc, \tag{6.43}$$

from which it is easy to derive the generalized Ward identity (6.14). Differentiating the equality (6.43) with respect to η and assuming $\bar{\eta}, \eta = 0$, we have

$$0 = \int \exp\left\{ i \int [\mathcal{L}_{\vartheta} + J_{\mu}^{a}A_{\mu}^{a}]\,dx \right\} \left\{ -\frac{1}{\alpha}\partial_{\mu}A_{\mu}^{a}(y) + \right.$$
$$\left. + \int \bar{c}^{a}(y)\,J_{\mu}^{b}(z)\,[\nabla_{\mu}c(z)]^{b}\,dz \right\} \prod_{x} d\mathcal{A}\,d\bar{c}\,dc. \tag{6.44}$$

Performing integration over $\bar{c}$, c by means of the formula (6.17), we obtain exactly the identity (6.14). In an analogous way it is possible to obtain the generalized Ward identities for the case when the Yang–Mills field interacts with the fields of matter.

Until now we have considered only covariant α-gauges, which are usually dealt with in practical calculations. However, all the reasoning automatically applies to the more general case when the term fixing the gauge has the following form:

$$B(A) = \exp\left\{ \frac{i}{2} \int \Phi^{2}(\mathcal{A}, x)\,dx \right\}, \tag{6.45}$$

where $\Phi(\mathcal{A})$ is a functional of $\mathcal{A}(x)$ which in principle can involve, in addition to terms linear in $\mathcal{A}$, terms of higher order as well. In this case, in accordance with the general quantization procedure described in the third chapter, the operator M figuring in the generating functional for the Green functions is given by the formula (1.1.26):

$$M_\Phi \alpha = \int \frac{\delta \Phi(\mathscr{A}, x)}{\delta A_\mu(y)} \nabla_\mu \alpha(y)\, dy. \tag{6.46}$$

To obtain the generalized Ward identities for this case it is sufficient, in all the computations given at the beginning of the present section, to substitute

$$\begin{aligned} \partial_\mu A_\mu &\to \Phi(\mathscr{A}), \\ M &\to M_\Phi. \end{aligned} \tag{6.47}$$

As a result, instead of (6.14) we obtain

$$\int \exp\left\{ i \int \left\{ S_\Lambda + \left[\frac{1}{2}\Phi^2(\mathscr{A}) + J^a_\mu A^a_\mu\right] dx \right\} \det M_\Phi \times \right.$$
$$\times \left\{ \Phi^a(A, y) + \int J^b_\mu(z) \left(\nabla^z_\mu M^{-1}_\Phi\right)^{ba}(z, y, A)\, dz \right\} \prod_x d\mathscr{A} = 0. \tag{6.48}$$

4.7. THE STRUCTURE OF THE RENORMALIZED ACTION

Let us analyze the structure of the primitively divergent diagrams in the Yang–Mills theory. We shall start with the Yang–Mills field in vacuum. In the α-gauge the effective Lagrangian has the form

$$\mathscr{L} = \frac{1}{2} \operatorname{tr} \left\{ \frac{1}{4} [(\partial_\nu \mathscr{A}_\mu - \partial_\mu \mathscr{A}_\nu) + g[\mathscr{A}_\mu, \mathscr{A}_\nu]]^2 - \right.$$
$$\left. - \frac{1}{2\alpha} (f(\Box)\, \partial_\mu \mathscr{A}_\mu)^2 + \bar{c}\, \Box c - g\bar{c}\partial_\mu [\mathscr{A}_\mu, c] \right\}. \tag{7.1}$$

The diagram technique involves the following elements:

1. Vector lines $\overline{\mathscr{A}_\mu \mathscr{A}_\nu}$, lines of fictitious c particles $\overline{\bar{c}c}$. To these lines there correspond the free Green functions $D_{\mu\nu}(p)$ and $D(p)$, behaving asymptotically as p^{-2} as $p \to \infty$.
2. Vertices with three outgoing vector lines and one derivative.
3. Vertices with four vector lines and without derivatives.
4. Vertices with a single vector and two fictitious lines and one derivative.

In accordance with the general formula derived in Section 4.2, the index of a diagram containing n_3 three-legged vector vertices, n_4 four-legged vertices, n_c vertices with fictitious particles participating, L_{in} internal vector lines and L^c_{in} internal fictitious lines is equal to

$$\omega = 2L_{\text{in}} + 3L^c_{\text{in}} - 4(n_4 - 1) - 3(n_3 + n_c). \tag{7.2}$$

Taking advantage of the fact that the number of internal lines (L_{in}, L^c_{in}) is related to the number of external lines (L_{ex}) by

$$L_{\text{in}} = \frac{4n_4 + 3n_3 + n_c - L_{\text{ex}}}{2}; \tag{7.3}$$

$$L^c_{\text{in}} = \frac{2n_c - L^c_{\text{ex}}}{2},$$

we express the diagram index in terms of the number of external lines

$$\omega = 4 - L_{\text{ex}} - L^c_{\text{ex}}. \tag{7.4}$$

From this formula it follows that the perturbation-theory series for the Yang–Mills field in the α-gauge contains a finite number of types of primitively divergent diagrams. These diagrams are symbolically presented in Fig. 9. [Formally there exist also a logarithmically divergent diagram with two external vectors and two fictitious lines, and a divergent diagram with four fictitious lines. However, as is seen from the formula (7.1), the derivative in one of the vertices can be transposed to

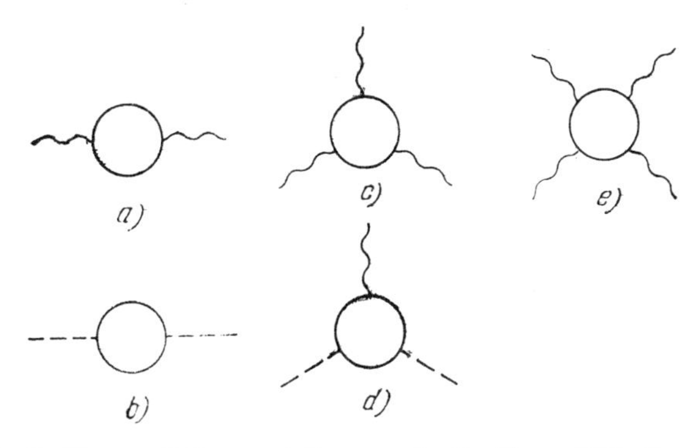

Fig. 9. Types of divergent diagrams in the Yang–Mills theory.

the external c-line by integrating by parts. Therefore in reality the corresponding diagrams converge.]

The self-energy diagrams in (a) and (b) have an index 2. The index of the diagrams in (c) and (d) is 1, and of the diagram in (e), 0. For the same reasons as above, the actual index of diagrams having external c-lines is reduced by 1. Besides this, for reasons of Lorentz invariance all diagrams with indices equal to one in reality diverge only logarithmically.

In accordance with the general procedure for the removal of these divergences it is necessary to subtract from the corresponding vertex functions several of the leading terms of the Taylor series in the external momenta. As the expansion center, a point is usually chosen at which the external momenta are on the mass shell, since such a choice provides for the proper normalization of one-particle states. However, in the case when the system under consideration involves particles of zero mass, the vertex functions on the mass shell may involve additional singularities due to the divergence of the corresponding integrals at the origin of the coordinate system (the infrared catastrophe). Therefore we shall perform subtractions at the points for which the values of all the external momenta are in the Euclidean region. For a vertex with n external momenta p_i, for example, such points are

$$p_i^2 = -a^2; \quad p_i p_j = \frac{a^2}{n-1}. \tag{7.5}$$

At these points all the vertex functions are real and are free from infrared singularities.

We shall now write out the most general expressions for the subtracted terms compatible with the conditions of relativistic invariance and of Bose symmetry. The proper vertex functions corresponding to the diagrams presented in Figure 9 have the following structure:

$$\begin{aligned}
\Gamma_{\mathcal{A}\mathcal{A}}(p) &\equiv \Gamma^{ab}_{\mu\nu}(p) = \delta^{ab}\{b_1 g_{\mu\nu} + b_2 p_\mu p_\nu + \\
&\qquad + b_3(p^2 g_{\mu\nu} - p_\mu p_\nu)\} + \dots, \\
\Gamma_{\bar{c}c}(p) &\equiv \Gamma^{ab}(p) = \delta^{ab} b_4 p^2 + \dots, \\
\Gamma_{\mathcal{A}^3} &\equiv \Gamma^{lmn}_{\lambda\nu\rho}(p, k, q) = i\varepsilon^{lmn} b_5 \{g_{\lambda\nu}(p-k)_\rho + \\
&\qquad + g_{\nu\rho}(k-q)_\lambda + g_{\lambda\rho}(q-p)_\nu\} + \dots, \\
\Gamma_{\mathcal{A}\bar{c}c} &\equiv \Gamma^{lmn}_{\mu}(p, k, q) = \frac{i}{2}\varepsilon^{lmn} b_6 (k-q)_\mu + \dots,
\end{aligned}$$

$$\Gamma_{\mathcal{A}^4} \equiv \Gamma^{ilmn}_{\mu\nu\rho\lambda}(p,\, k,\, q,\, r) =$$
$$= P\,\{b_7 \varepsilon^{gil} \varepsilon^{gmn} g_{\mu\rho} g_{\nu\lambda} + b_8 \delta^{il} \delta^{mn} g_{\mu\nu} g_{\lambda\rho}\} + \ldots \tag{7.6}$$

(Here the vertex functions for the group SU_2 have been written down. In this case the tensor structure in the charge indices is limited to the tensors ε^{abc} and δ^{ab}. In the general case there may be present additional linearly independent structures, such as, for example, terms proportional to the symmetric tensor d^{abc} in the case of the group SU_3. The proof of the renormalizability given below remains valid in this case too.)

In the last formula P is the symmetrization operator of the pairs of indices $(i,\, \mu)$, $(l,\, \nu)$, $(m,\, \rho)$, $(n,\, \lambda)$. The cofficients b_i depend on the regularization parameters Λ, μ and on the positions of the subtraction points a_i; $\cdots$ denotes the subsequent terms of the Taylor expansion, which tend to a finite limit when the regularization is removed. The subtraction of the polynomials (7.6) is equivalent to the insertion in the effective Lagrangian of the following counterterms:

$$\begin{aligned}\Delta\mathcal{L} = \frac{1}{4}\operatorname{tr}\Big\{ & b_1 \mathcal{A}_\mu^2 + b_2 (\partial_\mu \mathcal{A}_\mu)^2 + \frac{b_3}{2} (\partial_\nu \mathcal{A}_\mu - \partial_\mu \mathcal{A}_\nu)^2 + \\ & - 2b_4 \bar{c}\,\Box c + b_5 (\partial_\nu \mathcal{A}_\mu - \partial_\mu \mathcal{A}_\nu)[\mathcal{A}_\mu,\, \mathcal{A}_\nu] + \\ & + b_6 \bar{c}\partial_\mu [\mathcal{A}_\mu,\, c] + \frac{b_7}{2} [\mathcal{A}_\mu,\, \mathcal{A}_\nu][\mathcal{A}_\mu,\, \mathcal{A}_\nu]\Big\} + \\ & + \frac{b_8}{16} \{\operatorname{tr}(\mathcal{A}_\mu \mathcal{A}_\mu)\}^2.\end{aligned} \tag{7.7}$$

The number of types of counterterms needed for the removal of divergences is finite, and therefore the theory is renormalizable. However, the counterterm part of the action (7.7) contains significantly more parameters than the original Lagrangian. For arbitrary parameters b_i the renormalized theory is not gauge-invariant and does not satisfy the relativity principle, and that leads to the loss of the equivalence of various gauges and as a consequence to the violation of unitarity.

When the intermediate regularization is fixed, then the values of the parameters b_i depend on the choice of the subtraction points a_i. We shall show that this arbitrariness allows one to choose the parameters b_i so that the renormalized theory becomes gauge-invariant.

Let us ascertain what constraints are imposed by the relativity principle on the form of the renormalized effective Lagrangian. First of all note that the gauge-invariance of the Yang–Mills Lagrangian is not violated if it is multiplied by a constant. In addition we can treat arbitrarily the parameter g, which plays the part of the charge. In other words, the most general expression for the gauge-invariant Lagrangian of the Yang–Mills field has the form

$$\mathscr{L} = \frac{z_2}{8} \operatorname{tr} \{(\partial_\nu \mathscr{A}_\mu - \partial_\mu \mathscr{A}_\nu) + g z_1 z_2^{-1} [\mathscr{A}_\mu, \mathscr{A}_\nu]\}^2. \qquad (7.8)$$

The role of the parameter of the gauge transformation for this Lagrangian is played by the constant $\tilde{g}$:

$$\tilde{g} = g z_1 z_2^{-1}. \qquad (7.9)$$

The same parameter must be involved in the definition of the covariant derivative to provide for the self-consistency of the theory. Specifically, in the operator M, which has been defined by the formula

$$M = \partial_\mu \nabla_\mu = \partial_\mu (\partial_\mu - g [\mathscr{A}_\mu, \ \ ]), \qquad (7.10)$$

the constant $\tilde{g}$ must be substituted for g. Writing $\det M(\tilde{g})$ in the form of an integral over the fields of fictitious particles,

$$\det M(\tilde{g}) = \text{const} \int \exp \left\{ \frac{-i \operatorname{tr}}{2} \int \tilde{z}_2 \bar{c} \partial_\mu (\partial_\mu c - \tilde{z}_1 \tilde{z}_2^{-1} g \times \right. \\ \left. \times [\mathscr{A}_\mu, c]) \, dx \right\} \prod_x d\bar{c} \, dc, \qquad (7.11)$$

where

$$\tilde{z}_2^{-1} \tilde{z}_1 = z_2^{-1} z_1, \qquad (7.12)$$

and we have again taken the opportunity to multiply the Lagrangian as a whole by an arbitrary constant, we obtain the most general expression for the admissible effective Lagrangian:

$$\mathscr{L}_R = \frac{1}{2} \operatorname{tr} \left\{ \frac{1}{4} z_2 [(\partial_\nu \mathscr{A}_\mu - \partial_\mu \mathscr{A}_\nu) + z_1 z_2^{-1} g [\mathscr{A}_\mu, \mathscr{A}_\nu]]^2 - \right. \\ \left. - \frac{1}{2\alpha} (f(\Box) \partial_\mu \mathscr{A}_\mu)^2 - \tilde{z}_2 (\bar{c} \Box c - \tilde{z}_1 \tilde{z}_2^{-1} g \bar{c} \partial_\mu [\mathscr{A}_\mu, c]) \right\}. \qquad (7.13)$$

The constants z_1, z_2, $\tilde{z}_1$, $\tilde{z}_2$ are related by (7.12). The familiar condition $z_1 = z_2$ is not necessary and, generally speaking, does not hold: the Lagrangian (7.13) has the same structure as the nonrenormalized Lagrangian, differing from the latter only by the multiplicative renormalization of the fields $\mathscr{A}_\mu$, $\bar{c}$, c of the charge g and of the gauge parameter α:

$$\mathscr{A}_\mu \to z_2^{1/2}\mathscr{A}_\mu \quad c \to \tilde{z}_2^{1/2} c, \quad \bar{c} \to \tilde{z}_2^{1/2}\bar{c}, \tag{7.14}$$

$$g \to z_1 z_2^{-3/2} g, \quad \alpha \to z_2 \alpha.$$

Unlike the general expression (7.7), this Lagrangian involves only three independent counterterms, and at first sight it is not obvious that with their aid it is possible to remove all the divergences.

By introducing an invariant intermediate regularization it is possible to construct a generating functional for the Green functions $Z_R(J)$ corresponding to the Lagrangian (7.13). We shall assume that the regularization is performed, using the method of higher covariant derivatives decribed in Section 4.3. Since in all further reasoning only the invariance of the regularized action will be used, we shall not write the regularizing terms explicitly; they are described, for example, by the formula (4.34).

The role of the parameter of the gauge transformation for the Lagrangian (7.13) is played by the constant $\tilde{g} = z_1 z_2^{-1} g$. Therefore, the generalized Ward identitites which are satisfied by the functional $Z_R(J)$ differ from the identities (6.15) by the substitution of $\tilde{g}$ for g in the operators M^{-1} and ∇_μ:

$$\frac{1}{\alpha} f^2(\Box)\,\partial_\mu^x \left\{ \frac{1}{i} \frac{\delta Z_R}{\delta J_\mu^a(x)} \right\} + \left\{ \int J_\mu^b(y) \left[\delta^{bd}\partial_\mu^y - \tilde{g} t^{bcd} \frac{1}{i} \frac{\delta}{\delta J_\mu^c(y)} \right] \times \times \tilde{M}_{yx}^{-1\,da} \left(\frac{1}{i} \frac{\delta}{\delta J} \right) dy \right\} Z_R = 0, \tag{7.15}$$

where the sign $\sim$ means that the constant g involved in the definition of the operator M^{-1} is replaced by $\tilde{g}$.

In this formula it is convenient to pass to the renormalized Green functions of the fictitious particles, defined by the equality

$$G_R^{da}(y, x, J) = N^{-1} \int \bar{c}^d(y)\, c^a(x) \exp\left\{ i \int \left[\tilde{z}_1 \bar{c}^i(s) \Box c^i(s) - - \tilde{z}_1 t^{ikl} g \bar{c}^i(s)\, \partial_\mu \left[A_\mu^k(s)\, c^l(s) \right] + \mathscr{L}_R^{YM}(s) + + A_\mu^a J_\mu^a \right] ds \right\} \prod_x d\mathscr{A}\, d\bar{c}\, dc. \tag{7.16}$$

For this, note that

$$\tilde{M}_{yx}^{-1ab} Z_R = N^{-1} \int \bar{c}^a(y) c^b(x) \exp\Big\{ i \int [\bar{c}^a(s) \square c^a(s) - \\ - \tilde{z}_2^{-1} \tilde{z}_1 g t^{abd} \bar{c}^a(s) \partial_\mu [A_\mu^b(s) c^d(s)] + \mathscr{L}_R^{YM}(s) + \\ + A_\mu^a J_\mu^a]\, ds \Big\} \prod_x d\mathscr{A}\, d\bar{c}\, dc = \tilde{z}_2 G_R^{ab}(y, x, J). \quad (7.17)$$

The last relation is obtained from (7.16) by the substitution of variables

$$c \to \tilde{z}_2^{-1/2} c, \quad \bar{c} \to \tilde{z}_2^{-1/2} \bar{c}. \quad (7.18)$$

Representing the source J_μ on the right-hand side of the identity (7.17) in the form $J_\mu = J_\mu^{tr} + \partial_\mu \square^{-1} \partial_\nu J_\nu$, and taking advantage of the fact that

$$\partial_\mu^y (\nabla_\mu(\tilde{g}) \tilde{M}^{-1})_{yx}^{ab} = \delta^{ab} \delta(x - y), \quad (7.19)$$

we rewrite these relations in the form

$$\frac{1}{\alpha} f^2(\square) \partial_\mu^x \left\{ \frac{1}{i} \frac{\delta Z_R}{\delta J_\mu^a(x)} \right\} = \\ = \left\{ \int D_0(x - y) \partial_\mu J_\mu^a(y)\, dy \right\} Z_R + \\ + \int J_\mu^{tr\, b}(y) g \tilde{z}_1 t^{bcd} \frac{1}{i} \frac{\delta}{\delta J_\mu^c(y)} G_R^{da}(y, x, J)\, dy. \quad (7.20)$$

We shall show that for a suitable choice of the constants $z_2, \tilde{z}_2, \tilde{z}_1$, the finiteness of all the Green functions follows from the identity (7.20). The proof will be by induction. Assuming all the diagrams up to and including the nth order to be finite, we shall show that the functional F on the right-hand-side of the equation (7.20) is finite to the order $n + 1$. Hence it follows that the functional

$$\partial_\mu^x \left\{ \frac{1}{i} \frac{\delta Z_R}{\delta J_\mu^a(x)} \right\} \quad (7.21)$$

is also finite to order $n + 1$. This, as we shall see, means that all the Green functions (except, maybe, the two-point functions of the Yang–

Mills field Γ_{AA} and of the fictitious particles $\Gamma_{\bar{c}c}$) are finite. The divergences of these functions are removed by using the renormalization constants z_2 and $\tilde{z}_2$, the choice of which remains at our disposal.

The proof is particularly simple in the case of gauges for which the longitudinal part of the Green function of the Yang–Mills field decreases rapidly at large momenta, that is, when the function $f(k^2)$ involved in the definition of the generalized α-gauge behaves asymptotically as k^{2n} for $n > 1$. [This, of course, includes also the Lorentz gauge itself ($\alpha = 0$) when the longitudinal part of $D_{\mu\nu}$ is equal to zero.]

The proof for an arbitrary gauge does not involve any notions that are new in principle, but it is more cumbersome. So, not to distract the reader with technical details, we shall first consider the case

$$f(k^2) \xrightarrow[k^2 \to \infty]{} k^{2n}, \quad n \geqslant 1,$$

and come back to the discussion of more general gauges later, when we are investigating the dependence of the Green functions on the gauge.

In the lowest order of g^2 only the two-point functions Γ_{AA} and $\Gamma_{\bar{c}c}$ diverge. The proper vertex function $\Gamma_{AA} \equiv \Gamma_{\mu\nu}^{mn}(k)$ in second order is connected with the Green function $G_{\mu\nu}^{ab}(k)$ by the relation

$$G_{\alpha\beta}^{ab}(k) = D_{\alpha\mu}^{am}(k)\,\Gamma_{\mu\nu}^{mn}(k)\,D_{\nu\beta}^{nb}(k) + D_{\alpha\beta}^{ab}(k). \qquad (7.22)$$

From the identity (7.20) it follows that

$$\frac{f^2(-k^2)}{\alpha} k^\alpha k^\beta G_{\alpha\beta}^{ab}(k) = \frac{\alpha k_\mu}{k^2} \Gamma_{\mu\nu}^{ab}(k^2) \times$$
$$\times \frac{k_\nu}{k^2 f^2(-k^2)} + \delta^{ab} = \delta^{ab}, \qquad (7.23)$$

that is, the function $\Gamma_{\mu\nu}^{ab}(k)$ is transverse:

$$\Gamma_{\mu\nu}^{ab}(k) = \delta^{ab}\left(k^2 g_{\mu\nu} - k_\mu k_\nu\right) \Pi\left(k^2\right). \qquad (7.24)$$

Therefore the constants b_1 and b_2 in the Lagrangian (7.7) are equal to zero, and for the removal of the divergence one counterterm $z_2^{(0)}$ is sufficient. In an analogous way the counterterm $\tilde{z}_2^{(0)}$ provides for the finiteness of the Green function of the fictitious particles.

We shall now prove the finiteness of the third-order vertex functions. To the vertex function $\Gamma_{\bar{c}c\mathcal{A}}$ there correspond the strongly connected diagrams presented in Figure 10. As was shown above, the index of these diagrams equals zero, which means they are formally

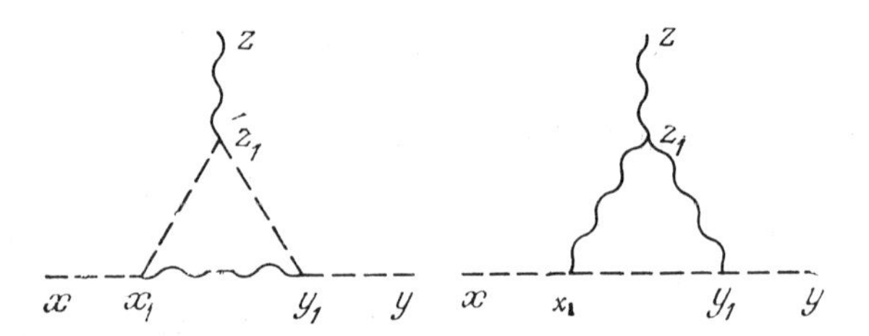

Fig. 10. Third-order corrections to the vertex function $\Gamma_{\bar{c}cA}$.

logarithmically divergent. It is not difficult to verify, however, that in the gauges for which

$$\tilde{f}(k^2)\xrightarrow[k\to\infty]{} |k|^{2n}, \quad n \geqslant 1,$$

the divergence is absent. Indeed, the analytical expression corresponding to the diagrams in Figure 10, represents in the momentum representation the sum of terms of the type

$$\sim \int D(x-x_1)\,\partial_\nu^{x_1}\{D(x_1-z_1)\,\partial_\rho^{z_1}[D_{\rho\mu}(z_1-z)\,D(z_1-y_1)]\times \\ \times \partial_\lambda^{y_1}[D_{\nu\lambda}(x_1-y_1)\,D(y_1-y)]\}\,dx_1\,dy_1\,dz_1.$$

Integrating by parts, it is easy to transform this expression so that the derivatives at the vertices x_1 and y_1 act either on the vector Green function, as a result of which only its rapidly decreasing longitudinal part gives a contribution to the integral, or on the external line. Therefore the true index of divergence is reduced, and convergent integrals correspond to the diagrams in Figure 10. Correspondingly, in the gauges under consideration the constant $\tilde{z}_1$ is finite.

For the Green function $G^{abc}_{\mu\nu\rho}(x, y, z)$ the identity (7.20) gives

$$\frac{1}{\alpha} f^2(\Box)\,\partial_\mu^x G^{abc}_{\mu\nu\rho}(x, y, z) = g\tilde{z}_1 t^{bed}\left(g_{\nu\alpha} - \frac{\partial_\nu\partial_\alpha}{\Box}\right)\times \\ \times \frac{1}{(i)^3}\left.\frac{\delta^2 G^{da}_R(y, x, J)}{\delta J^e_\alpha(y)\,\delta J^c_\rho(z)}\right|_{J=0} + (b\leftrightarrow c,\ y\leftrightarrow z,\ \nu\leftrightarrow\rho). \quad (7.25)$$

To the functions

$$\left.\frac{\delta^2 G^{da}_R(y, x, J)}{\delta J^e_\alpha(y)\,\delta J^c_\rho(z)}\right|_{J=0}$$

there correspond the diagrams in Fig. 11. The diagrams in (b) and (c) are weakly connected, and the convergence of the corresponding integrals follows from the finiteness of the second-order two-point Green functions $G^{ab}_{\mu\nu}$ and G^{ab}.

The diagram in (a) has a structure analogous to the structure of the diagram describing the transition of two fictitious particles into one vector particle (see Figure 10). It differs from the latter only in the form of the extreme left-hand vertex, denoted in the figure by a small cross. Only one vector and one fictitious line leave this vertex, while derivatives are absent. Formally the diagram in Figure 11(a) has a zero index, but for the same reasons as above the actual index is smaller by unity, and the divergence is absent.

Thus, the right-hand side of the equality (7.25) is finite, and consequently the function $\partial^x_\mu G^{abc}_{\mu\nu\rho}(x, y, z)$ is also finite in the third order of $g \cdots$. Therefore the divergence of the upper vertex function $\Gamma^{abc}_{\mu\nu\rho}$ is also finite:

$$(k+p)_\mu \Gamma^{abc}_{\mu\nu\rho}(k, p) < \infty. \tag{7.26}$$

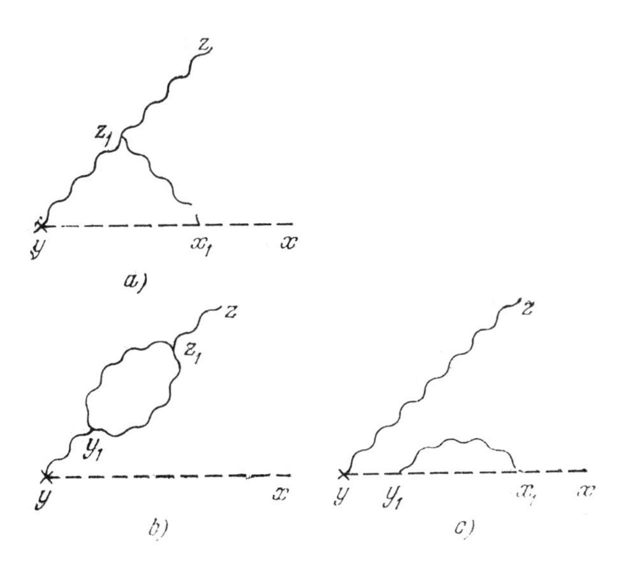

Fig. 11. Diagrams corresponding to the function $\left.\dfrac{\delta^2 G^{da}_R(y, x, J)}{\delta J^e_\alpha(y)\delta J^c_\rho(z)}\right|_{J=0}$.

The divergent part of $\Gamma^{abc}_{\mu\nu\rho}(k, p)$ can only be a polynomial of order not higher than one. The condition (7.26) signifies that this polynomial is identically equal to zero, and consequently, the function $\Gamma^{abc}_{\mu\nu\rho}(k, p)$ is finite in the third order of g. For the removal of the divergence from the vertex function $\Gamma^{abc}_{\mu\nu\rho}$ we have not had to introduce the independent renormalization constant z_1. This function turns out to be finite automatically if $z_1 = \tilde{z}_1\tilde{z}_2^{-1}z_2$.

The proof of the finiteness of the vertex functions of arbitrary order is absolutely analogous.

Let us consider the functional F in the right-hand side of Equation (7.20). Its connected part is represented by the diagrams in Figure 12. Let all the diagrams involved in the expansion of this functional be finite in all orders up to n. To prove the finiteness of the functional F in the $(n+1)$th order, it is sufficient to consider the diagrams in Figure 12 in which insertions in the external lines are absent, all the subgraphs being assumed to be finite.

The diagrams in Figure 12 are analogous to the diagrams representing Green functions $G(x, y, J)$ with two external fictitious lines and an arbitrary number of external vector lines. They differ only in the form of the vertex denoted by a small cross. One vector and one fictitious line leave this vertex, while derivatives are absent.

Therefore, the index of the diagrams in Figure 12 is the same as that of the diagrams corresponding to $G(x, y, J)$. The diagram with two external lines has an index equal to one. For reasons of Lorentz invariance the corresponding analytical expression has the form

$$\int J^{\text{tr}}_{\mu}(y)\,\Phi_{\mu}(y-x)\,dy = \int J^{\text{tr}}_{\mu}(y)\,\partial_{\mu}\Phi(y-x)\,dy = 0.$$

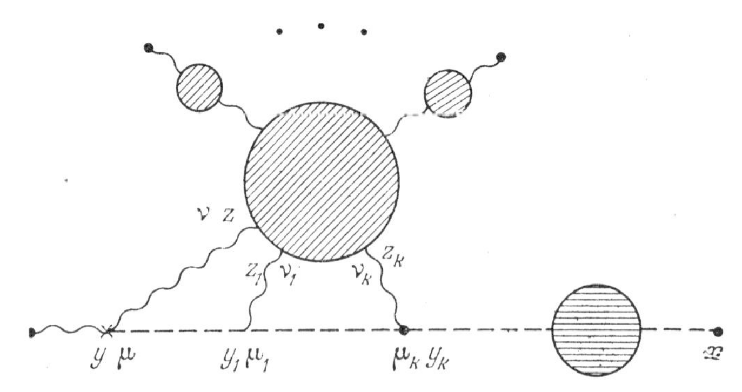

Fig. 12. ● ~ denotes an external classical source J_μ.

The diagram with three external lines has a zero index and in principle diverges logarithmically. The remaining diagrams all have negative indices. Repeating word for word the reasoning given above for the third-order diagrams, we come to the conclusion that logarithmic divergences in the diagrams in Figure 12 and also in the diagrams responsible for the vertex function $\Gamma_{\bar{c}c\mathscr{A}}$ are indeed absent.

Thus the assumption that the integrals corresponding to diagrams of the n-th order are convergent leads to the finiteness of the functional F in all orders up to $n+1$ inclusively. This means that all the integrals involved in the expansion of the functional in the left-hand side of the equation (7.20) also converge, and thus all the Green functions are finite:

$$\partial^{x_1}_{\mu_1}\left\{\frac{1}{i}\frac{\delta}{\delta J^{a_1}_{\mu_1}(x_1)}\cdots\frac{1}{i}\frac{\delta}{\delta J^{a_m}_{\mu_m}(x_m)}\right\}Z_R = \\ = \partial^{x_1}_{\mu_1}G^{a_1\,\cdots\,a_m}_{\mu_1\,\cdots\,\mu_m}(x_1\,\dots\,x_m). \qquad (7.27)$$

The Fourier transforms of the function $G^{a_1\,\cdots\,a_m}_{\mu_1\,\cdots\,\mu_m}$ are related to the vertex functions $\Gamma^{a_1\,\cdots\,a_m}_{\mu_1\,\cdots\,\mu_m}$ by the relation

$$G^{a_1\,\cdots\,a_m}_{\mu_1\,\cdots\,\mu_m}(k_1\,\dots\,k_m) = \\ = G^{a_1b_1}_{\mu_1\nu_1}(k_1)\,\dots\,G^{a_mb_m}_{\mu_m\nu_m}(k_m)\,\Gamma^{b_1\,\cdots\,b_m}_{\nu_1\,\cdots\,\nu_m}(k_1\,\dots\,k_m). \qquad (7.28)$$

All the two-point functions $G^{a_ib_i}_{\mu_i\nu_i}$ are reversible and are of order $\leq n$ in g. Therefore from the finiteness of the functions (7.27) it follows that

$$k_{\mu_1}\Gamma^{a_1\,\cdots\,a_m}_{\mu_1\,\cdots\,\mu_m}(k_1\,\dots\,k_m) = \varphi(k_1\,\dots\,k_m) < \infty. \qquad (7.29)$$

Generally speaking, both strongly and weakly connected diagrams correspond to the vertex functions $\Gamma^{a_2\,\cdots\,a_m}_{\mu_1\,\cdots\,\mu_m}$. To weakly connected diagrams there correspond coefficient functions representable (in the momentum representation) in the form of a product of coefficient functions of a lower order, which are assumed to be finite. Therefore, the equality (7.29) can be considered to be correct for the proper vertex functions of the order $n+1$.

If the proper vertex function has an index 0 or 1, then its divergent part can only be a polynomial of order not higher than one. The condition (7.29) signifies that this polynomial is identically equal to zero, and consequently, all the functions $\Gamma^{a_1 \cdots a_m}_{\mu_1 \cdots \mu_m}(k_1 \cdots k_m)$ are finite. The only possible exceptions are the two-point Green functions of the Yang–Mills field and of the field of the fictitious c-particles. The latter is not even involved in the expansion (7.27) and therefore is not subject to any limitations.

With regard to the two-point Green function of the Yang–Mills fields, since the diagram corresponding to it has an index 2, the divergent part of $\Gamma^{a_1 a_2}_{\mu_1 \mu_2}(k)$ can be a second-order polynomial. The condition (7.29) is insufficient for turning a second-order polynomial into zero. It does not impose any restrictions on its transverse part

$$\text{const}\,(g_{\mu\nu}k^2 - k_\mu k_\nu). \tag{7.30}$$

We have, however, at our disposal two more arbitrary counterterms of order $n+1$: $z_2^{(n+1)}, \tilde{z}_2^{(n+1)}$. These counterterms are sufficient for the removal of divergences from the two-point Green functions. Thus, all the diagrams of the $(n+1)$th order are finite. The induction has come to an end.

Now let the Yang–Mills field interact also with scalar and spinor fields. The corresponding Lagrangians are given by the formulas (1.3.1, 11). The diagram technique, besides the elements already discussed, contains now scalar and spinor lines, to which there correspond the Green functions $D(p)$ and $S(p)$ with the asymptotic behavior p^{-2} and p^{-1}, respectively, vertices with two spinor lines and one vector line without derivatives, vertices with two scalar lines and one vector line with one derivative, and vertices with two vector and two scalar lines without derivatives. The index of the diagram with L^A_{ex} external vector, L^c_{ex} fictitious, L^φ_{ex} scalar, and L^ψ_{ex} spinor external lines is equal to

$$\omega = 4 - L^A_{\text{ex}} - L^c_{\text{ex}} - L^\varphi_{\text{ex}} - \frac{3}{2} L^\psi_{\text{ex}}. \tag{7.31}$$

Besides the diagrams already mentioned, the diagrams presented in Figure 13 also have a nonnegative index. The self-energy diagram for the scalar field in (c) diverges quadratically, the diagrams in (a), (d) linearly, and the remaining diagrams logarithmically.

The corresponding proper vertex functions have the form [again for definiteness we write the formulas for the case of the group $SU(2)$]

$$\begin{aligned}
\Gamma_{\bar{\psi}\psi} &= (d_1 + \gamma_\mu p_\mu d_2)\,\delta^{ab} + \ldots,\\
\Gamma_{\varphi\varphi} &= (d_3 + d_4 p^2)\,\delta^{ab} + \ldots,\\
\Gamma_{\bar{\psi}\psi A} &= d_5 \gamma_\mu \varepsilon^{abc} + \ldots,\\
\Gamma_{\varphi\varphi A} &= i d_6 \varepsilon^{abc} (k - q)^\mu + \ldots,\\
\Gamma_{\varphi\varphi AA} &= d_7 g_{\mu\nu} (\delta^{ab}\delta^{cd} - \delta^{ac}\delta^{bd}) +\\
&\qquad + d_8 g_{\mu\nu} (\delta^{ab}\delta^{cd} + \delta^{ac}\delta^{bd}) + \ldots,\\
\Gamma_{\varphi^4} &= d_9 (\delta^{ab}\delta^{cd} + \delta^{ac}\delta^{bd} + \delta^{ad}\delta^{bc}) + \ldots, \qquad (7.32)
\end{aligned}$$

where $\cdots$ stands for the terms tending to a definite limit when the regularization is removed. As in the case of the Yang–Mills field in vacuum, the number of possible counterterms is significantly greater than the number of parameters in the nonrenormalized Lagrangians.

The most general expression for the gauge-invariant renormalized Lagrangian is constructed as before, and has the form

$$\begin{aligned}
\mathscr{L}^R = \mathscr{L}^R_{YM} &+ i z_{2\psi} \bar{\psi} \gamma_\mu \left(\partial_\mu - z_{2\psi}^{-1} z_{1\psi} g A^a_\mu \Gamma(T^a)\right) \psi -\\
&- z_{2\psi} (m + d)\, \bar{\psi}\psi + \frac{1}{2} z_{2\varphi} \left(\partial_\mu \varphi - z_{1\varphi} z_{2\varphi}^{-1} g A^a_\mu \Gamma(T^a)\, \varphi\right)^2 -\\
&- \frac{z_{2\varphi}}{2} (m^2 + f)\, \varphi^2 + z\lambda\, (\varphi^2)^2, \qquad (7.33)
\end{aligned}$$

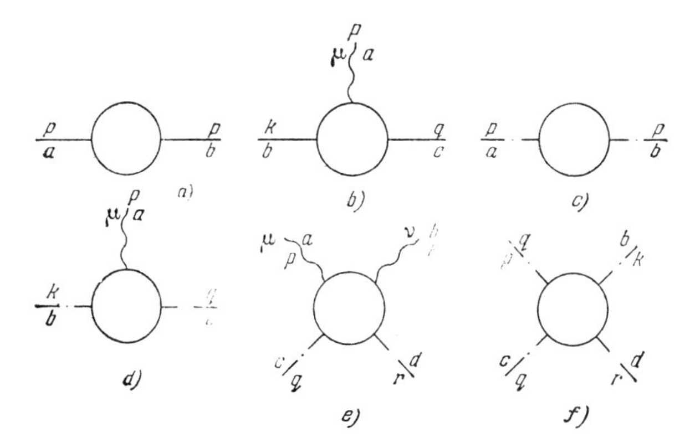

Figure 13. Additional divergent diagrams in the theory of the Yang–Mills field interacting with spinor and scalar fields. The solid line denotes the propagator of a spinor particle; the dash-dotted line, of a scalar particle.

where $\mathscr{L}^R_{YM}$ is the renormalized Lagrangian of the Yang–Mills field in vacuum (7.13).

Gauge invariance requires that the constants $z_{2\psi}^{-1}z_{1\psi}g$ and $z_{2\varphi}^{-1}z_{1\varphi}g$ and involved in the covariant derivatives of the spinor and scalar fields, coincide with the corresponding constant $\tilde{g} = z_2^{-1}z_1g$ which figures in the Lagrangian of the Yang–Mills field:

$$z_{2\psi}^{-1}z_{1\psi} = z_{2\varphi}^{-1}z_{1\varphi} = z_2^{-1}z_1. \tag{7.34}$$

As before, the conditions $z_{2\psi} = z_{1\psi}$, $z_{2\varphi} = z_{1\varphi}$ are not necessary. The gauge invariance does not impose any restrictions on the counter-terms d and f, renormalizing the masses of the fields and the counter-term $z\lambda(\varphi^2)^2$. We shall choose the constants $z_{2\psi}$ and $z_{2\varphi}$ in accordance with the condition of finiteness of the two-point Green functions of the spinor and scalar fields.

If the condition (7.34) is fulfilled, then the Green functions generated by the Lagrangian (7.33) satisfy the generalized Ward identities (6.26) with the obvious substitution $g \to \tilde{g}$. The proof of the finiteness of the Green function repeats word for word the reasoning given above. The only difference is that the functional F on the right-hand side of (7.20) contains additional terms

$$\tilde{z}_1 g \int \zeta^b(y)\,(\Gamma^d)^{bc}\left[\frac{1}{i}\,\frac{\delta G^{da}_{yx}}{\delta\zeta^c(y)}\right]dy + \ldots, \tag{7.35}$$

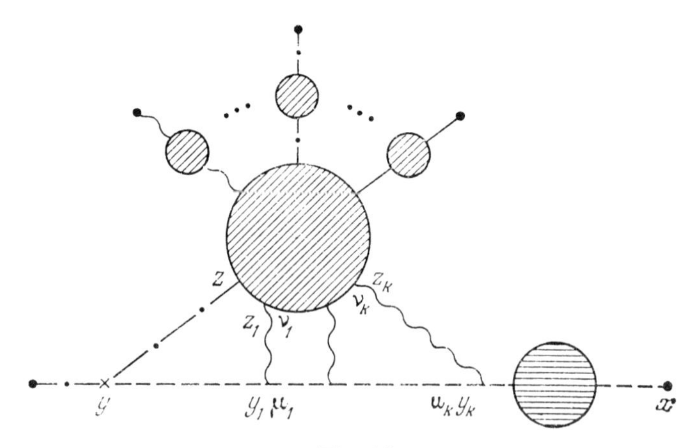

Fig. 14.

where $\cdots$ denotes analogous terms for the spinor fields. The corresponding diagrams are presented in Figure 14. The finiteness of these diagrams is demonstrated exactly in the same manner as the finiteness of the diagrams in Figure 12.

All the remaining reasoning is entirely identical to the analogous reasoning for the Yang–Mills field in vacuum. Thus, in order to remove all the ultraviolet divergences the gauge-invariant counterterms are sufficient in this case also.

Nor do any new features appear in the theory with spontaneously broken symmetry. The above-described scheme for proving the renormalizability remains unchanged. Consider, for example, the model (1.3.25). The most general form of an admissible renormalized Lagrangian can be obtained in the following manner. In accordance with the procedure described above, admissible counterterms

$$\mathcal{L}' = z_{2\varphi}\left|\left(\partial_\mu\varphi^+ - iz_{1\varphi}z_{2\varphi}^{-1}g\,\frac{\tau^a}{2}\,A_\mu^a\varphi^+\right)\right|^2 - \\ - z\lambda^2(\varphi^+\varphi - \mu^2 + \delta\mu^2)^2 + \mathcal{L}_{YM}^R. \quad (7.36)$$

are inserted in the Lagrangian (1.3.25). The constants $z_{2\varphi}$, $z_{1\varphi}$ satisfy the conditions (7.34). Passing to the shifted fields B^a, σ defined by the formula (6.31), we obtain

$$\mathcal{L}_R = \frac{z_{1\varphi}^2 z_{2\varphi}^{-1}}{2}\,m_1^2 A_\mu^2 + z_{1\varphi}m_1 A_\mu^a\partial_\mu B^a + \frac{1}{2}\,z_{2\varphi}\partial_\mu B^a\partial_\mu B^a + \\ + \frac{1}{2}\,z_{2\varphi}\partial_\mu\sigma\partial_\mu\sigma - \frac{m_2^2}{2}\,\sigma^2 - \frac{zm_2^2}{m_1^2}\frac{g^2}{8}\,\delta\mu^2(B^2+\sigma^2) + \\ - \frac{zm_2^2}{m_1}\frac{g}{2}\,\delta\mu^2\sigma + z_{1\varphi}\,\frac{g}{2}\,A_\mu^a\left(\sigma\partial_\mu B^a - B^a\partial_\mu\sigma - \varepsilon^{abc}B^b\partial_\mu B^c\right) + \\ + z_{1\varphi}^2 z_{2\varphi}^{-1}\left[\frac{m_1 g}{2}\,\sigma A_\mu^2 + \frac{g^2}{8}\,(\sigma^2+B^2)\,A_\mu^2\right] - \\ - \frac{zgm_2^2}{4m_1}\,\sigma(\sigma^2+B^2) - \frac{zg^2m_2^2}{32m_1^2}\,(\sigma^2+B^2)^2. \quad (7.37)$$

The renormalized Yang–Mills Lagrangian (7.13), involving also interaction with fictitious particles, remains unchanged, and we shall not write it out.

In passing to the formula (7.37) we performed a shift of the fields φ by a quantity equal to the vacuum expectation value of the field φ, not allowing for radiative corrections. Therefore in the Lagrangian (7.37) there are present counterterms linear in the field σ, which compensate the divergences in the "tadpole"-type diagrams (Figure 15), and also the counterterm renormalizing the masses of the Goldstone fields B^a. These counterterms are neccesary in order to provide for the equilibrium of the ground state when taking the radiative corrections into account.

The Lagrangian (7.36) is invariant with respect to the gauge transformations (6.32) with the substitution

$$m_1 \to \tilde{m}_1, \quad g \to \tilde{g}, \quad \tilde{g} = \tilde{z}_1 \tilde{z}_2^{-1} g, \quad \tilde{m}_1 = \tilde{z}_1 \tilde{z}_2^{-1} m_1. \tag{7.38}$$

The generalized Ward identities are modified in the same way as in the symmetric case

$$\begin{aligned} \frac{1}{\alpha} f^2(\Box)\, \partial_\mu \left[\frac{1}{i} \frac{\delta Z_R}{\delta J_\mu^a(x)}\right] = \int \{D_0(x-y)\, \partial_\mu J_\mu^a(y)\, dy\} Z_R + \\ + \int \Bigg\{ (J_\mu^b)^{\text{tr}}(y)\, \tilde{z}_1 g \varepsilon^{bcd} \frac{1}{i} \frac{\delta}{\delta J_\mu^c(y)} + \frac{\tilde{z}_1 g}{2} \delta^{bd} \xi_\sigma(y) \frac{1}{i} \frac{\delta}{\delta \xi_B^b(y)} - \\ - \xi_B^b(y)\, \tilde{z}_1 \left[\frac{g}{2} \varepsilon^{bcd} \frac{1}{i} \frac{\delta}{\delta \xi_B^c(y)} + \delta^{bd} \frac{g}{2} \frac{1}{i} \frac{\delta}{\delta \xi_\sigma(y)} + m_1 \delta^{bd}\right] \Bigg\} \times \\ \times G_R^{da}(J, \xi_B, \xi_\sigma, y, x)\, dy \end{aligned} \tag{7.39}$$

(we recall that in the gauges under consideration the constant $\tilde{z}_1$ is finite). The proof of the renormalizability repeats practically word for word the corresponding reasoning for the symmetric case. The only technical complication consists in that due to the mixing of the fields A_μ^a, B^a in the generalized gauge, the two-point Green functions are represented by matrices (2×2).

The last comment concerns the normalization of the two-point Green functions. Since in this case we deal with massive particles, the subtractions may be performed on the mass shell. Therefore we shall assume the counterterms to be chosen so that the poles of the total two-point Green function of the σ-particles and of the transverse part of the Green function of the Yang–Mills field coincide with the poles of the corresponding free functions. We cannot, however, in general provide simultaneously for the residues at the corresponding poles to be equal to unity. As is seen from the formula (7.38), the counterterms renormalizing the masses and the wave functions are not independent. Having defined the position of the pole of the Green function, we are not free to handle the value of the residue at this pole arbitrarily. Therefore, to

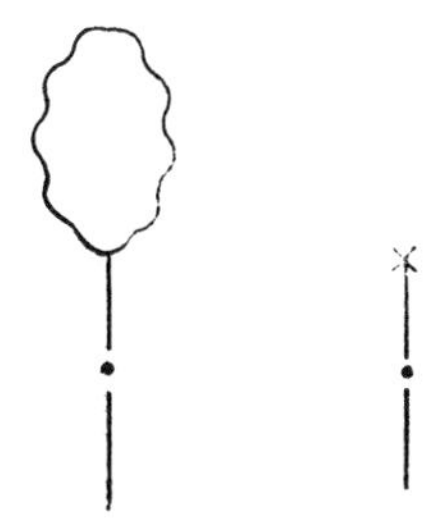

Fig. 15. Diagrams of the "tadpole" type in the Yang–Mills theory with spontaneously broken symmetry.

provide for the correct normalization of the single-particle states it is necessary to perform an additional finite renormalization.

4.8. THE RENORMALIZED S-MATRIX

We have shown that the renormalization procedure can be carried out without violating the gauge invariance of the theory. We shall now show that the renormalized theory obeys the relativity principle, meaning that the probabilities of physical processes do not depend on the actual choice of the gauge condition. Thus the unitarity of the renormalized S-matrix will be proved.

In the present section we shall consider models with spontaneously broken symmetry, in which all physical particles have nonzero masses. Formally all the reasoning may be applied to the symmetric theory also, but in this case, as has already been pointed out, the matrix elements on the mass shell contain additional infrared singularities. Therefore, in the framework of perturbation theory, the S-matrix in the symmetric theory, strictly speaking, does not exist.

So let us consider the renormalized generating functional for the Green functions, which can be written as

$$Z(J_\mu, \zeta_\sigma) = N^{-1} \int \exp\left\{ i \int [\mathscr{L}_R + J^a_\mu A^a_\mu + \zeta_\sigma \sigma]\, dx \right\} \times \\ \times \Delta(\mathscr{A}) \prod_x \delta(\partial_\mu \mathscr{A}_\mu)\, d\mathscr{A}\, d\sigma\, d\mathscr{B}. \tag{8.1}$$

Here $\mathscr{L}_R$ is the renormalized gauge-invariant Lagrangian of the Yang–Mills field interacting with the fields of matter. [For definiteness we consider the Lagrangian (7.36)]. We shall assume the source J^a_μ to be transverse:

$$\partial_\mu J^a_\mu = 0. \tag{8.2}$$

The matrix elements of the S-matrix are expressed in terms of the variational derivatives Z by means of reduction formulas

$$\begin{aligned} & S_{i_1 \dots i_n, j_1 \dots j_m}(k'_1 \dots k'_n, p'_1 \dots p'_l; k_1 \dots \\ & \dots k_m, p_1 \dots p_q) V^{\frac{n+m}{2}} W^{\frac{l+q}{2}} = \\ & = (k'^2_1 - m^2_1) \dots (k'^2_n - m^2_1)(p'^2_1 - m^2_2) \dots \\ & \dots (p'^2_l - m^2_2)(k^2_1 - m^2_1) \dots (k^2_m - m^2_1)(p^2_1 - m^2_2) \dots \\ & \dots (p^2_q - m^2_2)\theta(k'_{10}) \dots \theta(k'_{n0})\theta(-k_{10}) \dots \theta(-k_{m0})\theta(p'_{10}) \dots \\ & \dots \theta(p'_{l0})\theta(-p_{10}) \dots \theta(-p_{q0}) \times \\ & \times u^{i_1}_{\mu_1} \dots u^{i_n}_{\mu_n} G_{\mu_1 \dots \mu_n \nu_1 \dots \nu_m}(k'_1 \dots p_q) \times \\ & \times u^{j_1}_{\nu_1} \dots u^{j_m}_{\nu_m} \Big|_{p^2 = p'^2 = m^2_2;\ k^2 = k'^2 = m^2_1}. \end{aligned} \tag{8.3}$$

Here k, k' denote the momenta of the vector particles, and p, p' those of the scalar particles. The constants V and W are renormalizing factors:

$$\begin{aligned} & \delta^{ab}\left(g_{\mu\nu} - \frac{k_\mu k_\nu}{k^2}\right) \cdot V \underset{k^2 = m^2_1}{=} \\ & = (k^2 - m^2_1)\left(g_{\mu\nu} - \frac{k_\mu k_\nu}{k^2}\right) \int e^{ikx} \frac{\delta^2 Z}{\delta J^a_\mu(x)\, \delta J^b_\nu(0)}\Big|_{J,\, \zeta = 0} dx, \end{aligned} \tag{8.4}$$

$$W \underset{p^2 = m^2_2}{=} (p^2 - m^2_2) \int e^{ipx} \frac{\delta^2 Z}{\delta \zeta_\sigma(x)\, \delta \zeta_\sigma(0)}\Big|_{J,\, \zeta = 0} dx. \tag{8.5}$$

If the two-point Green function is normalized on the mass shell to unity,

$$(p^2 - m^2)\, G(p^2) = 1, \qquad p^2 = m^2, \tag{8.6}$$

then these factors are absent, and we come back to the formula (III.3.64). The matrix elements (8.3), calculated in the Lorentz gauge, tend to a definite limit when the intermediate regularization is removed. Let us demonstrate that in reality the values of the matrix elements (8.3) are independent of the choice of the gauge condition.

Let us pass in the expression (8.1) for the generating functional $Z(J, \zeta\ \sigma)$ from the Lorentz gauge to the unitary gauge

$$B^a = 0. \tag{8.7}$$

The invariance of the renormalized Lagrangian $\mathscr{L}_R$ allows us to use for this the same method as was used in Chapter 2.

By introducing the gauge-invariant functional $\triangle'(\sigma, B, \tilde{g})$ defined by the equality

$$\triangle'(\sigma, \mathscr{B}, \tilde{g}) \int \delta(\mathscr{B}^\omega) \prod_x d\omega = 1, \tag{8.8}$$

where

$$\mathscr{B}^\omega = \mathscr{B} - \tilde{m}_1 u - \frac{\tilde{g}}{2}[\mathscr{B}, u] - \frac{\tilde{g}}{2}\sigma u + O(u^2), \tag{8.9}$$

we can rewrite the functional $Z(J, \zeta\ )$ in the form

$$Z(J_\mu, \zeta) = N^{-1} \int \exp\left\{ i \int \left[\mathscr{L}_R + J^a_\mu A^a_\mu + \zeta_\sigma \sigma \right] dx \right\} \triangle(\tilde{g}, \mathscr{A}) \times$$
$$\times \prod_x \delta(\partial_\mu \mathscr{A}_\mu)\, \triangle'(\mathscr{B}, \tilde{g}, \sigma)\, \delta(\mathscr{B}^\omega)\, d\omega\, d\mathscr{A}\, d\sigma\, d\mathscr{B}. \tag{8.10}$$

Passing to new variables

$$\mathscr{A}_\mu \to \mathscr{A}_\mu^{\omega^{-1}}, \quad \mathscr{B} \to \mathscr{B}^{\omega^{-1}}, \quad \sigma \to \sigma^{\omega^{-1}}, \quad \omega^{-1} \to \omega \tag{8.11}$$

and integrating over ω, we obtain, in complete analogy with the results of Chapter 2,

$$Z(J, \zeta) = N^{-1} \int \exp\left\{ i \int \left[\mathscr{L}_R + J^a_\mu (\mathscr{A}^\omega_\mu)^a + \zeta_\sigma \sigma^\omega \right] dx \right\} \times$$
$$\times \triangle'(\sigma, \mathscr{B}, \tilde{g}) \prod_x \delta(\mathscr{B})\, d\mathscr{A}\, d\sigma\, d\mathscr{B}, \tag{8.12}$$

where

$$\mathscr{A}^\omega_\mu = \mathscr{A}_\mu + \delta\mathscr{A}_\mu = \mathscr{A}_\mu + \partial_\mu u - \tilde{g}[\mathscr{A}_\mu, u] + O(u^2),$$
$$\sigma^\omega = \sigma + \delta\sigma = \sigma - \frac{\tilde{g}}{2}(\mathscr{B}u) + O(u^2), \tag{8.13}$$

and the function u is defined by the equation

$$\partial_\mu \mathscr{A}^\omega_\mu = \Box u - \tilde{g}\partial_\mu [\mathscr{A}_\mu, u] + \partial_\mu \mathscr{A}_\mu + \ldots = 0. \quad (8.14)$$

The value of the functional $\triangle'(\sigma, \mathscr{B}, \tilde{g})$ on the surface $\mathscr{B} = 0$ is equal to

$$\triangle'(\sigma, \mathscr{B}, \tilde{g}) \underset{B=0}{=} \det \left| \tilde{m}_1 + \frac{\tilde{g}\sigma(x)}{2} \right|^3 = \text{const} \det \left| m_1 + \frac{g\sigma(x)}{2} \right|^3. \quad (8.15)$$

The functional (8.12) differs from the generating functional for the Green functions in the unitary gauge only in the form of the terms with sources. We shall now show that if it is substituted in the reduction formula (8.3), the this difference vanishes; that means that the renormalized matrix elements remain unchanged for the substitution

$$J^a_\mu (\mathscr{A}^\omega_\mu)^a \to J^a_\mu \mathscr{A}^a_\mu, \quad \zeta_\sigma \sigma^\omega \to \zeta_\sigma \sigma. \quad (8.16)$$

The variational derivatives of the functional (8.12) are expressed in terms of Green functions of the form

$$\left(\frac{1}{i}\right)^{m+q} \frac{\delta^{(m+q)} Z}{\delta J^{a_1}_{\mu_1}(x_1) \ldots \delta J^{a_m}_{\mu_m}(x_m)\, \delta\zeta(y_1) \ldots \delta\zeta(y_q)} \Bigg|_{J,\, \zeta=0} =$$

$$= N^{-1} \int \exp\left\{ i \int [\mathscr{L}_R]\, dx \right\} \triangle'(\sigma, \mathscr{B}, \tilde{g}) \left(\mathscr{A}^\omega_{\mu_1}\right)^{a_1}(x_1) \ldots$$

$$\ldots \left(\mathscr{A}^\omega_{\mu_m}\right)^{a_m}(x_m)\, \sigma^\omega(x_1) \ldots \sigma^\omega(x_q) \prod_x \delta(\mathscr{B})\, d\mathscr{A}\, d\sigma\, d\mathscr{B}, \quad (8.17)$$

where $\mathscr{A}^\omega_\mu$, σ^ω are defined by the formulas (8.13). Since the sources J_μ are considered to be transverse, the linear term $\partial_\mu u$ does not give any contribution and the perturbation-theory expansion of $\delta\mathscr{A}_\mu$ and $\delta\sigma$ start with terms quadratic in the fields.

The diagrams presented in Figure 16 correspond to the Green functions (8.17). Diagrams of the types in (a) and (b) contain poles in all variables p_i, k_j. Diagrams of the type in (c) are one-particle irreducible at least in one of the momenta p_i, k_j. (The diagram presented in the figure is one-particle irreducible in the momentum p_1. This means that it is not possible to split it into two parts, connected only by one line, along which the momentum p_1 propagates.)

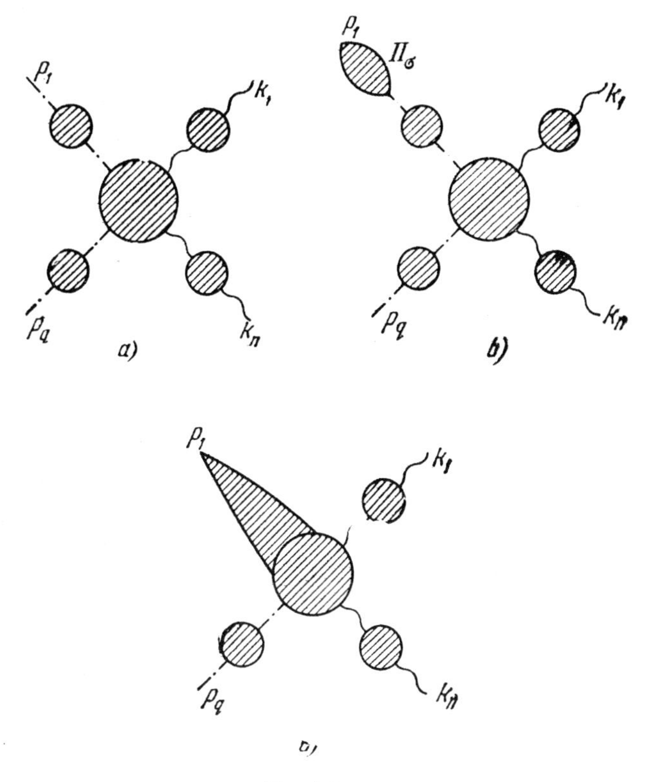

Fig. 16.

From the investigation of the analytical properties of Feynman diagrams it is known that one-particle irreducible diagrams have no pole singularities in the corresponding variables. Therefore if the coefficient functions corresponding to the diagram in (c) are multiplied by the product:

$$\prod_i (k_i^2 - m_1^2) \prod_j (p_j^2 - m_2^2) \tag{8.18}$$

and $k_1^2 = m_1^2$, $p_j^2 = m_2^2$ is assumed, then this expression vanishes.

Diagrams of the (b) type are obtained from (a)-type diagrams by means of insertions in the external lines of the blocks denoted in Figure 16 as $\Pi_{\mathcal{A}}$, Π_σ. On the mass shell this is equivalent to multiplying the corresponding Green functions by constants equal to the values of the functions $\Pi_A(k^2)$ and $\Pi_\sigma(p^2)$ at the points $k^2 = m_1^2$, $p^2 = m_2^2$. The values of the Green functions with m external vector and q external scalar lines on the mass shell when one gauge is changed to another are changed as follows:

$$\prod_{i=1}^{m}(k_i^2-m_1^2)\prod_{j=1}^{q}(p_j^2-m_2^2)\,G^{(u)\,a_1\ldots a_m}_{\mu_1\ldots\mu_m}(k_1,\ldots,k_m,p_1,\ldots,p_q)=$$
$$=(1+\Pi_{\mathscr{A}}(m_1^2)^n(1+\Pi_\sigma(m_2^2))^m\prod_{i=1}^{n}(k_i^2-m_1^2)\prod_{j=1}^{q}(p_j^2-m_2^2)\times$$
$$\times\,G^{(L)\,a_1\ldots a_m}_{\mu_1\ldots\mu_m}(k_1,\ldots,k_m,p_1,\ldots,p_q)\Big|_{\substack{k_i^2=m_1^2\\ p_j^2=m_2^2}}. \tag{8.19}$$

Here $G^{(u)}$ is the Green function in the unitary gauge, and $G^{(L)}$ in the Lorentz gauge. Obviously, a quite analogous formula relates the Green functions in other gauges also.

For the two-point Green function the corresponding transformation is presented graphically in Figure 17. The values of the two-point Green functions on the mass shell in various gauges are connected by the relation

$$(k^2-m_1^2)\,G^{(L)\,ab}_{\mu\nu}(k)\underset{k^2=m_1^2}{=}(1+\Pi_{\mathscr{A}}(m_1^2))^2(k^2-m_1^2)\,G^{(u)\,ab}_{\mu\nu}(k),$$
$$(p^2-m_2^2)\,G^{(L)}_\sigma(p)\underset{p^2=m_2^2}{=}(1+\Pi_\sigma(m_2^2))^2(p^2-m_2^2)\,G^{(u)}_\sigma(p). \tag{8.20}$$

Going back to the formula (8.3), we see that passing to the unitary gauge signifies that the Fourier transforms of the Green functions with m vector and q scalar external lines are multiplied by

$$(1+\Pi_{\mathscr{A}}(m_1^2))^m(1+\Pi_\sigma(m_2^2))^q, \tag{8.21}$$

but simultaneously the normalizing constants V and W are multiplied by $(1+\Pi_{\mathscr{A}}(m_1^2))^2$ and $(1+\Pi_\delta(m_2^2))^2$, respectively. As a result, the expression for the renormalized matrix element remains unchanged.

From this reasoning, which is actually the analog of the Borhers theorem in the axiomatic quantum theory, it follows that the renormalized S-matrix is independent of the concrete choice of the gauge condition, and consequently the renormalized theory satisfies the relativity principle.

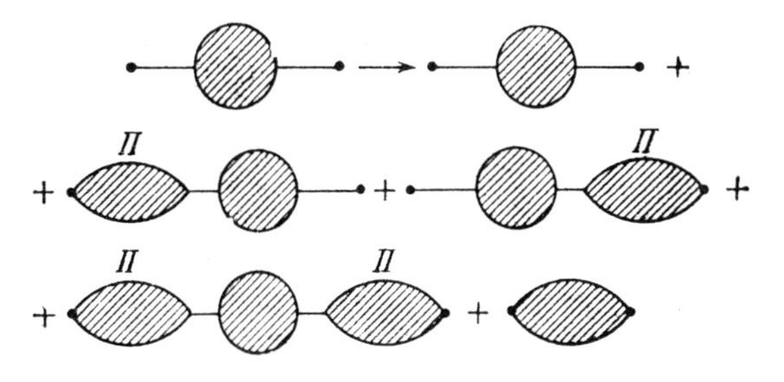

Fig. 17.

In the gauge $\mathscr{B} = 0$ the renormalized Lagrangian has the form

$$\mathscr{L}_R = \mathscr{L}^R_{YM} + \\ + \left\{ \frac{z_{1\varphi}^2 z_{2\varphi}^{-1}}{2} m_1^2 A_\mu^2 + \frac{1}{2} z_{2\varphi} \partial_\mu \sigma \partial_\mu \sigma - \frac{(m_2^2 + \delta m_2^2)\sigma^2}{2} - \frac{2\delta m_2^2 m_1}{g} \sigma + \right. \\ \left. + \frac{m_1 g}{2} z_{1\varphi}^2 z_{2\varphi}^{-1} \sigma A_\mu^2 + \frac{z_{1\varphi} z_{2\varphi}^{-1}}{8} g^2 \sigma^2 A_\mu^2 - \frac{zgm_2^2}{4m_1} \sigma^3 - \frac{zg^2 m_2^2}{32 m_1^2} \sigma^4 \right\}. \tag{8.22}$$

All nonphysical particles (the Goldstone bosons, the fictitious c-particles, the longitudinal quanta of the vector field) are absent, and the unitarity of the scattering matrix is obvious. Due to the independence of the S-matrix of the gauge, the matrix elements on the mass shell tend to a definite limit when the regularization is removed. Note that this is not true of the Green functions of the mass shell, generated by the Lagrangian (8.22). The free Green function of the vector field corresponding to the Lagrangian (8.22) has the form

$$D_{\mu\nu} = \frac{-1}{(2\pi)^4} \frac{g_{\mu\nu} - k_\mu k_\nu m_1^{-2}}{k^2 - m_1^2}, \tag{8.23}$$

and as $k \to \infty$ tends to a constant. Calculating the divergence index of the diagram containing n_3 trident vector vertices, n_4 four-legged vertices, and L_{ex} external vector lines, we find

$$\omega = 4 + 4n_4 + 2n_3 - 2L_{ex}; \tag{8.24}$$

with the increase of n_i the number of types of divergent diagrams increases infinitely, that is, off the mass shell the theory is non-renormalizable. Nevertheless for the removal of the divergences from the matrix elements on the mass shell a finite number of the counter-terms written out in the formula (8.22) are sufficient. Gauge-invariance leads to the physical equivalence between the explicitly renormalizable and explicitly unitary gauges, due to which the renormalized S-matrix has both these properties.

Obviously, these conclusions do not depend on the concrete model (8.22); they all apply equally to the model (1.3.13) and also to the models involving additional gauge-invariant interaction of fermions. Only the gauge invariance of the renormalized Lagrangian is essential.

To conclude this section we shall go back to the question of the proof of the finiteness of the Green functions in the generalized renormalizable gauge. Until now we have considered either the Lorentz gauge, or gauges in which the longitudinal part of the vector Green function decreases rapidly as $k \to \infty$. We shall now show that this condition is not necessary and that the counterterms of the form (7.3) provide for the finiteness of the Green functions in any renormalizable gauges, that is, in gauges for which the longitudinal part of the free Green function of the vector field decreases no more slowly than the transverse part as $k \to \infty$. The simplest example of such a gauge is the gauge with $f(k^2) = 1$.

In any such gauge the divergent diagrams have a structure already discussed above: only diagrams with one, two, three, and four external lines can diverge. As before, we can choose the constants z_2, $\tilde{z}_2$, $z_{2\varphi}$, $\tilde{z}_{2\psi}$, $\tilde{z}_1$, z, δm so as to make all the two-point Green functions and the vertex functions $\Gamma_{\bar{c}cA}$, Γ_{σ^4} finite, and determine the constants z_1, $z_{1\varphi}$, $z_{1\psi}$ by the invariance of the renormalized action:

$$z_1 z_2^{-1} = \tilde{z}_1 \tilde{z}_2^{-1} = z_{1\varphi} z_{2\varphi}^{-1} = z_{1\psi} z_{2\psi}^{-1}. \tag{8.25}$$

Let us show that ratios of the type

$$\frac{\Gamma_{A^3}}{(\Gamma_{AA})^{3/2}}, \quad \frac{\Gamma_{\bar{\psi}\psi A}}{(\Gamma_{\bar{\psi}\psi})(\Gamma_{AA})^{1/2}}, \quad \text{and so on,} \tag{8.26}$$

where all the external vectors legs are considered to be transverse, on the mass shell do not depend on the gauge. Indeed, from the formulas (8.19) and (8.20) it follows that a transfer from one gauge to another changes the functions under consideration in the following manner (we omit the tensor structure):

$$\Gamma_{A^3}(k_1, k_2, k_3) \to (1 + \Pi_A(m_1^2))^3 \Gamma_{A^3}(k_1, k_2, k_3),$$
$$\Gamma_{AA}(k) \to (1 + \Pi_A(m_1^2))^2 \Gamma_{AA}(k), \qquad k_i^2 = m_1^2. \tag{8.27}$$

By substituting these expressions into the formula (8.26), we verify the invariance of this ratio. According to the above demonstration, in the Lorentz gauge all functions are finite. The function Γ_{AA} is finite due to the choice of constants z_i. Therefore in any gauge the function $\Gamma_{A^3}(k_1, k_2, k_3)$ is finite at $k_i^2 = m_1^2$. Since in renormalizable gauges this function can diverge only logarithmically, the finiteness of Γ_{A^3} at any k_i follows.

The finiteness of all the remaining Green functions is demonstrated absolutely analogously. We emphasize that now we are considering Green functions off the mass shell. On the mass shell divergences are absent in all gauges, including the not explicitly renormalizable ones (that is to say, gauges for which the longitudinal part of the Green function of the vector field at large k behaves as k^{2n}, $n > -1$).

In the renormalizable gauge a finite number of invariant counterterms provide for the existence of the Green functions off the mass shell also. Then the concrete values of the counterterms depend, of course, on the choice of the gauge. Specifically, in the general gauge the constant $\tilde{z}_1$ is already not finite.

4.9. ANOMALOUS WARD IDENTITIES

In the construction of the unitary renormalized S-matrix we used the invariant intermediate regularization. The existence of the invariant regularized action allowed us to obtain the generalized Ward identities and, using them, to prove the physical equivalence of the unitary and Lorentz gauges. Generally speaking, it is not necessary to employ the invariant intermediate regularization. In principle, we could introduce an arbitrary intermediate regularization and try to choose the counterterms in such a manner that the renormalized Green functions would satisfy the generalized Ward identities. For this, if the regularization is noninvariant, noninvariant counterterms such as the photon mass renormalization in electrodynamics may be required.

In this case, in the regularized theory the relativity principle is violated, and its correctness in the limit when the intermediate regularization is removed requires a special investigation. It may turn out that whatever the choice of local counterterms, the renormalized Green functions will not satisfy the generalized Ward identities. This will lead to the nonequivalence of different gauges and the inconsistency of the theory. In this case the unitary renormalized S-matrix does not (at least in the framework of perturbation theory) exist.

In practice the indicated situation arises when the matrix γ_5 is involved in gauge transformations of the fermion fields. In this case both the above described methods for invariant regularization are inapplicable. In the framework of the dimensional regularization it turns out not to be possible to give a consistent definition of the matrix γ_5 for a space with arbitrary dimensionality. The regularization by means of the higher covariant derivatives still provides the finiteness of all multiloop diagrams; however, the invariant regularization of one-loop diagrams by means of the Pauli–Villars procedure in this case is impossible, since the mass terms for the fermion fields $\mu_j\bar{\psi}_j\psi_j$ violate the γ_5-invariance. Thus, for one-loop diagrams there is no γ_5-invariant regularization, and, as we see, for a number of gauge groups involving γ_5-transformations, the Green functions do not satisfy the generalized Ward identities.

As a simple example, consider a model with the $U(1)$ gauge group, described by the Lagrangian

$$\mathscr{L} = -\frac{1}{4}(\partial_\nu A_\mu - \partial_\mu A_\nu)^2 + i\bar{\psi}\gamma_\mu{}'(\partial_\mu - igA_\mu\gamma_5)\psi,$$
$$\gamma_5 = -i\gamma_0\gamma_1\gamma_2\gamma_3. \tag{9.1}$$

This Lagrangian is invariant under the Abelian gauge transformations

$$A_\mu(x) \to A_\mu(x) + \partial_\mu\lambda(x),$$
$$\psi(x) \to e^{ig\gamma_5\lambda(x)}\psi(x);$$
$$\bar{\psi}(x) \to \bar{\psi}(x)e^{ig\gamma_5\lambda(x)}, \tag{9.2}$$

and at first sight all the reasoning for the equivalence of various gauges can be equally applied to it. In the α-gauge the effective Lagrangian has the form

$$\mathscr{L}_{\mathfrak{s}} = \mathscr{L} + \frac{1}{2\alpha}(\partial_\mu A_\mu)^2, \tag{9.3}$$

where $\mathscr{L}$ is the gauge-invariant expression (9.1). The Lagrangian (9.3) is nondegenerate and describes not only transversely polarized quanta of the vector field, but also scalar quanta with zero spin.

One might take the Lagrangian (9.3) as a starting point and base the construction of a quantum theory on it. It is well known that such a theory would be inconsistent physically: the probability of events involving scalar quanta can take negative values. If, however, the Green functions generated by the Lagrangian (9.3) satisfy the Ward identities

$$\left\{\frac{1}{\alpha}\partial_\mu\left(\frac{\delta Z}{\delta J_\mu(x)}\right) - -Z\partial_\mu J_\mu(x) + ig\bar{\eta}(x)\gamma_5\frac{\delta Z}{\delta\bar{\eta}(x)} + ig\frac{\delta Z}{\delta\eta(x)}\gamma_5\eta(x)\right\} = 0, \quad (9.4)$$

where Z is a generating functional of the form

$$Z = N^{-1}\int \exp\left\{i\int[\mathscr{L}_э + J_\mu A_\mu + \bar{\eta}\psi + \bar{\psi}\eta]\,dx\right\} dA\,d\bar{\psi}\,d\psi, \quad (9.5)$$

then, as it is easy to show, the matrix elements of transitions between states involving transversely polarized quanta and states involving scalar quanta is equal to zero. This means that the S-matrix connecting "physical" transversely polarized states is unitary. (Strictly speaking, the S-matrix does not exist in the model considered, due to infrared divergences. It can be shown, however, that all the reasoning may be applied to the case when the vector field has a nonzero mass and the infrared divergences are absent.)

Formally, the identity (9.4) follows from the invariance of the Lagrangian (9.1) under the transformations (9.2). A specific case is the relation

$$\partial^{x_1}_{\mu_1}\frac{\delta^n \ln Z}{\delta J_{\mu_1}(x_1)\ldots\delta J_{\mu_n}(x_n)}\bigg|_{J,\ \eta=0} = 0, \qquad n > 2, \quad (9.6)$$

which demonstrates explicitly the absence of transitions between transversely and longitudinally polarized states. In reality we are interested not in the naive identities (9.4), which, strictly speaking, have no sense because of the divergent integrals involved in them, but in the corresponding relations for the renormalized Green functions. In electrodynamics, as also in the non-Abelian models discussed above, the Green functions satisfy generalized Ward identities which differ from the "naive" ones only by the renormalization of the charges and masses involved. This is not so in the model (9.1). The Green function with three external vector lines, corresponding to the diagram presented in Figure 18, does not satisfy the "naive" identities (9.4), no matter what local counterterms have been chosen. The identity (9.6) means that the Fourier transform of the three-point vertex function $\Gamma_{\mu\nu\alpha}(p, q)$, defined by the equality

$$\Gamma_{\mu\nu\alpha}(p, q) G_{\mu\mu'}(p) G_{\nu\nu'}(q) G_{\alpha\alpha'}(p+q) = \\ = \int e^{ipx} e^{iqy} \left(\frac{-i\delta^3 Z}{\delta J_{\mu'}(x) \delta J_{\nu'}(y) \delta J_{\alpha'}(0)} \right) dx\, dy, \quad (9.7)$$

must be transverse:

$$p_\mu \Gamma_{\mu\nu\alpha}(p, q) = q_\nu \Gamma_{\mu\nu\alpha}(p, q) = (p+q)_\alpha \Gamma_{\mu\nu\alpha}(p, q) = 0. \quad (9.8)$$

The explicit calculation of $\Gamma_{\mu\nu\alpha}(p, q)$, taking into account that the function $\Gamma_{\mu\nu\alpha}(p, q)$ is symmetric in the arguments (p, μ), (q, ν), $(-(p+q), \alpha)$, gives

$$i(p+q)_\alpha \Gamma_{\mu\nu\alpha}(p, q) = -\frac{g^3}{6\pi^2} \varepsilon^{\mu\nu\alpha\beta} p_\alpha q_\beta \neq 0. \quad (9.9)$$

Since the index of the diagram in Figure 18 is equal to unity, the function $\Gamma_{\mu\nu\alpha}(p, q)$ is defined to the approximation of a first-order polynomial in p and q. One might try to use this arbitrariness in order to set the right-hand side of the equality (9.9) equal to zero. It is easy to see, however, that this is impossible. The most general expression for a renormalized three-legged vertex function has the form

$$\tilde{\Gamma}_{\mu\nu\alpha}(p, q) = \Gamma_{\mu\nu\alpha}(p, q) + c_1 \varepsilon_{\mu\nu\alpha\beta} p_\beta + c_2 \varepsilon_{\mu\nu\alpha\beta} q_\beta, \quad (9.10)$$

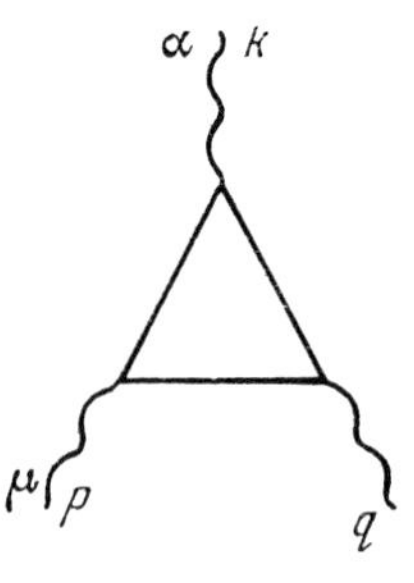

Fig. 18. Anomalous triangle diagram.

where $\Gamma_{\mu\nu\alpha}(p, q)$ is a symmetric vertex function satisfying the relation (9.9). By requiring the function $\tilde{\Gamma}_{\mu\nu\alpha}$ also to be symmetric in the arguments (μ, p), (ν, q), $(\alpha_1 - (p + q))$, we obtain

$$c_1 = c_2 = 0. \tag{9.11}$$

Thus, there is no possible choice of the local counterterms to make the renormalized vertex function satisfy the identity (9.8). In consequence, the probability of transition from transversely polarized states to longitudinal ones is not equal to zero. The model described by the Lagrangian (9.1) is inconsistent. The stated difficulty is inherent in all the theories invariant under the Abelian gauge transformations involving the matrix γ_5. However, there exists a class of models for which this difficulty can be avoided. For instance, in the model (9.1), suppose that besides the field ψ, a field ψ' is involved, which interacts with a vector field in the same way as ψ, but differs from the latter in the sign of the charge:

$$\mathcal{L} = -\frac{1}{4}(\partial_\nu A_\mu - \partial_\mu A_\nu)^2 + i\bar{\psi}\gamma_\mu(\partial - igA\gamma_5)\psi + \\ + i\bar{\psi}'\gamma_\mu(\partial + igA\gamma_5)\psi'. \tag{9.12}$$

Then together with the diagram in Figure 18 there is an analogous diagram, in which the internal lines correspond to the fields ψ'. From the formula (9.9) it may be seen that the divergence of the anomalous vertex function is proportional to g^3. Therefore, the diagram corresponding to ψ' will give the same contribution to the identity (9.9), but with opposite sign. As a result, the total vertex function $\Gamma_{\mu\nu\alpha}$ will satisfy the normal identities (9.8). By direct calculation it is not difficult to verify that all the remaining one-loop diagrams satisfy the identities (9.8). With regard to the diagrams containing more than one loop, the absence of anomalies can be proved for them in a general form. Indeed, as was shown in Section 4.3, regularization by means of the higher covariant derivatives makes all multiloop diagrams in any arbitrary gauge-invariant theory convergent. Therefore, if there are no anomalies in the one-loop diagrams, then the multiloop diagrams without doubt satisfy the normal Ward identities. The absence of anomalies in the model (9.12) can be explained also in the following way. It is possible to pass, in the Lagrangian (9.12), to new canonical variables

$$\psi_1 = \frac{1}{2}\{(1-\gamma_5)\psi + (1+\gamma_5)\psi'\};$$

$$\psi_2 = \frac{1}{2}\{(1+\gamma_5)\psi + (1-\gamma_5)\psi'\}. \tag{9.13}$$

The interaction Lagrangian expressed in terms of the fields ψ_1, ψ_2 does not contain the matrices γ_5,

$$\mathscr{L}_I = (g\bar{\psi}_1\gamma_\mu\psi_1 - g\bar{\psi}_2\gamma_\mu\psi_2) A_\mu, \tag{9.14}$$

and represents an analog of the electromagnetic interaction Lagrangian of two massless spinors. Such a theory is invariant under purely vector gauge transformations

$$\begin{aligned} \psi_1(x) &\to e^{ig\alpha(x)}\psi_1(x), \quad \bar{\psi}_1(x) \to e^{-ig\alpha(x)}\bar{\psi}_1(x), \\ \psi_2(x) &\to e^{-ig\alpha(x)}\psi_2(x), \quad \bar{\psi}_2(x) \to e^{ig\alpha(x)}\bar{\psi}_2(x), \\ A_\mu(x) &\to A_\mu(x) + \partial_\mu\alpha(x) \end{aligned} \tag{9.15}$$

and therefore the renormalized Green functions satisfy the normal Ward identities. Such a mechanism for compensating anomalies may be used also in more realistic models, specifically, in models with spontaneously broken symmetry. If the fields ψ, ψ', A_μ interact also with scalar fields, then with a corresponding choice of potential all physical particles may be made to acquire zero masses due to the Higgs mechanism. At the same time the form of interaction of the spinor and vector fields responsible for the appearance of anomalies remains unchanged. Therefore, all the reasoning concerning the compensation of anomalies remains correct.

Anomalous Ward identities may appear also in non-Abelian gauge fields. For example, let the spinor fields ψ interact in a gauge-invariant manner with the Yang–Mills field:

$$\mathscr{L} = i\bar{\psi}\gamma_\mu(\partial_\mu - g\Gamma^a A_\mu^a)\psi + \ldots, \tag{9.16}$$

and $\cdots$ stand for the Lagrangian of the Yang–Mills field and also, possibly, for the gauge-invariant interaction of the fields A_μ, ψ with scalar fields. The latter can correspond both to the symmetric theory and to the theory with spontaneously broken symmetry.

The matrices Γ^a realize the representation of the Lie algebra

$$[\Gamma^a, \Gamma^b] = t^{abc}\Gamma^c \tag{9.17}$$

and can also include the matrix γ_5. The divergence of the three-legged vertex Green function is calculated exactly as in the Abelian case. The only difference consists in the appearance of an additional factor, proprotional to the trace of the product of the Γ-matrices at the vertex:

$$i(p+q)_\alpha \Gamma^{abc}_{\mu\nu\alpha}(p, q) = \\ = \text{const} \operatorname{tr}\{\gamma_5 [\Gamma_a, \Gamma_b]_+ \Gamma_c\} \varepsilon_{\mu\nu\alpha\beta} p_\alpha q_\beta. \quad (9.18)$$

If the factor

$$A_{abc} = \operatorname{tr}\{\gamma_5 [\Gamma_a, \Gamma_b]_+ \Gamma_c\} \quad (9.19)$$

is not zero, then the function $\Gamma^{abc}_{\mu\nu\alpha}$ does not satisfy the generalized Ward identities, which leads to the loss of the gauge-invariance of the renormalized theory.

Let us analyze in which cases A_{abc} is equal to zero. For this, instead of the matrix Γ_α, we introduce the chiral matrices $T_\pm$

$$\Gamma_a = \frac{1}{2}(1+\gamma_5) T^+_a + \frac{1}{2}(1-\gamma_5) T^-_a, \quad (9.20)$$

where $T_\pm$ do not contain the matrix γ_5.

The factor A_{abc} can not be represented as

$$A_{abc} = 2(A^+_{abc} - A^-_{abc}), \quad (9.21)$$

where

$$A^\pm_{abc} = \operatorname{tr}\{[T^\pm_a, T^\pm_b]_+ T^\pm_c\}. \quad (9.22)$$

A_{abc} obviously becomes zero if $A^+ = A^-$. This is certainly fulfilled if the representations of $T^\pm_\alpha$ are unitarily equivalent

$$T^-_a = U T^+_a U^+, \quad (9.23)$$

where U is a unitary matrix. In this case, by another choice of the spinor-field basis, the interaction can be rewritten in a purely vector form:

$$\bar\psi\gamma_\mu\Gamma_a\psi = \frac{1}{2}\bar\psi\gamma_\mu\{(1+\gamma_5)T^+_a + (1-\gamma_5)T^-_a\}\psi = \\ = \bar\psi'\gamma_u T^+_a \psi', \quad (9.24)$$

where

$$\psi' = \frac{1}{2}(1 + \gamma_5)\psi + \frac{1}{2}(1 - \gamma_5)U\psi. \tag{9.25}$$

Under such a redefinition of the fields ψ the γ_5-matrices appear in the mass terms. The absence of the anomalies in such models is absolutely natural. In the basis ψ' the gauge transformations no longer contain the matrices γ_5, and therefore one can apply to them the above described procedure of invariant regularization which allowed us to prove the generalized Ward identities rigorously. The actual form of the gauge group is not essential.

Such models are called "vectorlike", since at high energies, exceeding significantly all characteristic masses, they behave as models with purely vector interaction.

The unitary equivalence of T_+ and T_- is not necessary for the absence of anomalies. For this the fulfilment of the equality (9.21), which may be satisfied for other reasons, is sufficient.

Anomalies are absent also if $A^+_{abc} = A^-_{abc} = 0$, as occurs in some gauge groups. The sufficient condition for this is the following: the representations realized by the matrixes $T^\pm_a$ must be real. (A representation is called real if it is unitarily equivalent to its complex conjugate.) In this case

$$\begin{aligned}\operatorname{tr}\{[T^\pm_a, T^\pm_b]_+ T^\pm_c\} &= \operatorname{tr}\{[(T^\pm_a)^*, (T^\pm_b)^*]_+ (T^\pm_c)^*\} = \\ &= -\operatorname{tr}\{[T^\pm_a, T^\pm_b]_+ T^\pm_c\} = 0.\end{aligned} \tag{9.26}$$

Such a situation is realized for the algebras

$$\begin{aligned}&SU(2);\ SO(N),\ N \geqslant 5,\ N \neq 6;\ S_p(2N),\ N \geqslant 3;\\ &G(2);\ E(4);\ E(7);\ E(8),\end{aligned} \tag{9.27}$$

all representations of which are real. For the algebra $SU(3)$ anomalies are absent only for the representations 8 and $3+\bar{3}$.

In non-Abelian theories one-loop diagrams with four external vector lines can also be anomalous. On the other hand, if in the given model all the one-loop diagrams satisfy the normal Ward identities, the multiloop diagrams are sure to be free from anomalies. This, as we have already pointed out, follows directly from the fact that the regularization by means of the higher covariant derivatives regularizes multiloop diagrams in any gauge theory, including those which contain γ_5-

transformations. Therefore, multiloop diagrams automatically satisfy the normal Ward identities.

The above classification of "normal" and "anomalous" theories equally concerns theories with spontaneously broken symmetry. In the anomalous case the unitary and the renormalized gauges correspond to physically nonequivalent theories. In the unitary gauge the theory is not renormalizable, and therefore makes no sense in the framework of perturbation theory. On the contrary, in a renormalizable gauge the perturbation theory is constructed without difficulty; however, the S-matrix is not unitary in the space of physical states. Thus, requiring the absence of anomalies imposes strict constraints on the possible gauge-invariant models.

CHAPTER 5

CONCLUSION: SOME APPLICATIONS

In this chapter we shall discuss possible applications of gauge theories to the description of elementary-particle interactions. The experimental situation both in weak and in strong interactions is changing very rapidly and today it is difficult to give preference to any concrete model. We shall therefore restrict ourselves to the description of the most characteristic features of gauge-invariant elementary-particle models, without attempting to reflect the latest trends in this field. The examples to be considered are educational and illustrative in character.

5.1 UNIFIED MODELS OF WEAK AND ELECTROMAGNETIC INTERACTIONS

Until recently electrodynamics was the only example of successful applications in elementary-particle physics of quantum field theory in general and gauge-invariant theories in particular. At the same time, it has been noticed for quite a while that weak and electromagnetic interactions have much in common. From experiments it is known that weak interactions involve vector currents. This leads to the idea that as in electrodynamics the interaction takes place through an exchange of vector particles, which have become known as intermediate bosons. Like the electromagnetic current, the weak current is conserved. And finally, the weak interaction is universal—the interaction is characterized by a single constant (if one neglects any effects due to a mixing of various fundamental particles).

All these properties receive a natural explanation if one assumes that weak and electromagnetic interactions are described by a gauge-invariant theory, in which the Yang–Mills field plays the role of the

L. D. Faddeev and A. A. Slavnov. Gauge Fields: Introduction to Quantum Theory, ISBN 0-8053-9016-2.

Copyright © 1980 by The Benjamin/Cummings Publishing Company, Advanced Book Program. All rights reserved. No part of this publication may be reproduced, stored in a retrieval system, or transmitted, in any form or by any means, electronic, mechanical photocopying, recording, or otherwise, without the prior permission of the publisher.

interaction carrier. However, unlike the long-range electromagnetic interaction, the weak interaction has a finite interaction range, and consequently the corresponding vector fields must be massive. A second difference is that the weak interaction does not conserve parity. These differences, which for a long time hindered the construction of a realistic unified theory of weak and electromagnetic interactions, can be explained successfully by means of the Higgs mechanism. Spontaneous symmetry breaking allows one to select an "electromagnetic" direction in the internal charge space. The corresponding vector meson remains massless and interacts with the parity-conserving current. The other mesons acquire a nonzero mass, and their interaction does not conserve parity.

Let us consider the simplest realization of these ideas. The choice of the gauge group is, to a great extent, arbitrary. The dimension of the group must be not less than three, since it must involve, at a minimum, the generators corresponding to the photon (1) and the intermediate vector mesons (2). Taking into account only the "light" leptons—the electron, the muon, and the corresponding neutrinos—the minimum number of generators is equal to four. Indeed, the charged weak current has the following structure:

$$j_e^+ = \bar{\nu}_e \gamma_\mu (1 + \gamma_5) e = \bar{\psi}_e \gamma_\mu (1 + \gamma_5) \tau^+ \psi_e, \tag{1.1}$$

where

$$\psi_e = \begin{pmatrix} \nu_e \\ e \end{pmatrix} \tag{1.2}$$

(the muon current has an analogous structure).

Therefore the matrices $(1 + \gamma_5)\tau^+$ and $(1 + \gamma_5)\tau^-$ are involved in the Lie algebra of the gauge group. The minimal Lie algebra containing these matrices consists of the generators

$$(1 + \gamma_5)\tau^+, \quad (1 + \gamma_5)\tau^-, \quad (1 + \gamma_5)\tau_3 \tag{1.3}$$

and corresponds to the $SU(2)$ group. This algebra does not contain a generator by means of which it could be possible to construct a parity-conserving electromagnetic current. The simplest algebra generating both the electromagnetic and the charged currents corresponds to the group $U(2)$ and contains four generators, one of which corresponds to the neutral weak current.

Precisely this group forms a basis for the Weinberg-Salam model, which will be discussed below.

By not confining oneself to the known "light" leptons it is possible, by introducing additional weakly interacting particles, to construct a unified model with correct charged and electromagnetic currents, remaining within the group $SU(2)$. Such a model was proposed by Georgi and Glashow. However, this model does not agree with recent experimental data, and we shall not discuss it.

The possibilities pointed out are the minimal ones. In the literature, many models based on more complicated gauge groups are discussed.

Historically, when the first unified models were being constructed, neither the neutral currents nor the heavy leptons were known experimentally. At present both these predictions have been confirmed experimentally.

Below we shall describe in detail a model based on the group $SU(2) \times U(1)$—the Weinberg–Salam model. In the Weinberg–Salam model the electron and the electron neutrino are united in an SU_2 doublet L and singlet R:

$$L = \frac{1}{2}(1 + \gamma_5)\begin{pmatrix} \nu_e \\ e \end{pmatrix}, \quad R = \frac{1}{2}(1 - \gamma_5)\, e. \tag{1.4}$$

This choice of the multiplets is due to the fact that the right-hand- and the left-hand-polarized leptons are involved in the weak interaction nonsymmetrically—the right-hand-polarized neutrino is not observed experimentally. The muon and the muon neutrino are united in analogous multiplets. Henceforth we shall limit ourselves to the consideration of the electron sector. By requiring the weak charged currents to have a $V - A$ structure and the photon to interact only with the vector current of charged particles, we come to the following transformation law:

$$\begin{aligned} &L(x) \to L(x) - ig\frac{\tau^a}{2}\zeta^a(x)\, L(x) - \frac{ig_1}{2}\eta(x)\, L(x) + \ldots, \\ &R(x) \to R(x) - ig_1\eta(x)\, R(x) + \ldots \end{aligned} \tag{1.5}$$

Since the group $SU(2) \times U(1)$ is not simple, the gauge transformations involve two arbitrary parameters g and g_1. To the subgroups $SU(2)$ and $U(1)$ there correspond the following gauge fields: the isovector field A^a_μ and the singlet B_μ.

The gauge-invariant Lagrangian describing the interaction of the multiplets R and L with the Yang–Mills fields has the form

$$\mathcal{L} = \frac{1}{8} \operatorname{tr} \mathcal{F}_{\mu\nu}\mathcal{F}_{\mu\nu} - \frac{1}{4} G_{\mu\nu}G_{\mu\nu} + \\ + \bar{L} i\gamma_\mu \left[\partial_\mu + ig\frac{\tau^a}{2} A^a_\mu + \frac{ig_1}{2} \mathcal{B}_\mu\right] L + \bar{R} i\gamma_\mu \left[\partial_\mu + ig_1\mathcal{B}_\mu\right] R, \tag{1.6}$$

where $\mathcal{F}_{\mu\nu}$ is the strength tensor of the Yang–Mills field, and $G_{\mu\nu}$ is the analogous tensor for the Abelian field:

$$G_{\mu\nu} = \partial_\nu B_\mu - \partial_\mu B_\nu. \tag{1.7}$$

Note that the mass term

$$m(\bar{L}R + \bar{R}L) \tag{1.8}$$

for leptons is forbidden by the requirement of invariance under the transformations (1.5).

All the fields involved in the Lagrangian (1.6) have zero masses. However, if the fields $\mathcal{A}_\mu$, B_μ and R, L interact also with scalar fields, then they may also acquire nonzero masses due to the Higgs effect. Since all vector mesons, with the exception of the photon, must become massive, we shall take advantage of the concrete spontaneous-symmetry-breaking model (1.3.25). We introduce the complex doublet

$$\varphi = \begin{pmatrix} \varphi_1 \\ \varphi_2 \end{pmatrix}, \tag{1.9}$$

which transforms under the gauge transformations in the following way:

$$\varphi \to \varphi - ig\zeta^a(x)\frac{\tau^a}{2}\varphi - \frac{ig_1}{2}\eta(x)\varphi(x). \tag{1.10}$$

The gauge-invariant Lagrangian describing the interaction of φ with the fields $\mathcal{A}_\mu$, B_μ, R, L has the form

$$\mathcal{L} = \left|\partial_\mu\varphi + ig\frac{\tau^a}{2}A^a_\mu\varphi + \frac{ig_1}{2}B_\mu\varphi\right|^2 - \\ - G\{(\bar{L}\varphi)R + \bar{R}(\varphi^+ L)\} + \frac{m^2}{2}\varphi^2 - \lambda^2(\varphi^2)^2. \tag{1.11}$$

As we already know, an interaction of the type (1.11) generates spontaneous symmetry breaking: the vacuum expectation value of the field φ differs from zero, and to construct the perturbation theory around the asymmetric ground state it is necessary to pass to the shifted fields

$$\varphi \to \varphi' = \begin{pmatrix} \varphi_1 \\ \varphi_2 + r \end{pmatrix}; \quad (\operatorname{Im} r = 0). \tag{1.12}$$

As a result of this shift, mass terms for the vector fields appear:

$$\frac{r^2}{4}\left[g^2(A_\mu^1)^2 + g^2(A_\mu^2)^2 + (g_1 B_\mu - gA_\mu^3)^2\right]. \tag{1.13}$$

The diagonalization of the quadratic form (1.13) leads to the following spectrum of masses:

The charged mesons $W_\pm$,

$$W_{\mu\mp} = \frac{A_\mu^1 \pm A_\mu^2}{\sqrt{2}}, \tag{1.14}$$

acquire masses

$$m_W = \frac{1}{\sqrt{2}}\, gr. \tag{1.15}$$

The neutral mesons

$$Z_\mu = (g^2 + g_1^2)^{-1/2}(-gA_\mu^3 + g_1 B_\mu) \tag{1.16}$$

and

$$A_\mu = (g^2 + g_1^2)^{-1/2}(g_1 A_\mu^3 + gB_\mu) \tag{1.17}$$

acquire the masses $(r/\sqrt{2})(g^2 + g_1^2)^{1/2}$ and 0, respectively. As a result of the shift (1.12), the leptons also acquire nonzero masses. The mass term has the form

$$-G\left\{\bar{L}\begin{pmatrix} 0 \\ r \end{pmatrix} R + \bar{R}(0,\ r)L\right\} = -Gr\bar{e}e. \tag{1.18}$$

The neutrino remains massless.

Finally, using the expansion

$$\varphi_1 = \frac{1}{\sqrt{2}}(iB_1 + B_2); \quad \varphi_2 = r + \frac{1}{\sqrt{2}}(\sigma - iB_3), \tag{1.19}$$

we find that the field σ acquires a mass $2\lambda r$.

The Goldstone fields B_i have a zero mass and, as usual, can be removed by a gauge transformation.

The interaction of the leptons with the vector fields has the form

$$\mathscr{L}_I = \frac{-g}{2\sqrt{2}} \bar{\nu}_e \gamma_\mu (1+\gamma_5) e W^+_\mu +$$

$$\text{Hermitian conjugate} + \frac{g g_1}{(g^2+g_1^2)^{1/2}} \bar{e}\gamma_\mu e A_\mu +$$

$$+ \frac{(g^2+g_1^2)^{1/2}}{4} \left\{ \bar{\nu}_e \gamma_\mu (1+\gamma_5) \nu_e - 2\bar{e}\gamma_\mu \left[\gamma_5 + \frac{g^2 - 3g_1^2}{g^2+g_1^2} \right] e \right\} Z_\mu. \tag{1.20}$$

From this formula it is seen that the electromagnetic constant e and the Fermi weak-interaction constant G_F are expressed in terms of the parameters g and g_1 in the following way:

$$e = \frac{g g_1}{\sqrt{g^2+g_1^2}}, \tag{1.21}$$

$$\frac{G_F}{\sqrt{2}} = \frac{g^2}{8m_W^2}. \tag{1.22}$$

From (1.21) it follows that

$$e \leqslant g, \tag{1.23}$$

whence

$$m_W = \left(\frac{g^2\sqrt{2}}{8G_F} \right)^{1/2} \geqslant \left(\frac{e^2\sqrt{2}}{8G_F} \right)^{1/2} = 37.5 \text{ GeV}, \tag{1.24}$$

that is, the mass of the charged intermediate meson has a lower bound and is large .

An analogous estimate of the mass of the neutral meson gives

$$M_Z \geq 75 \text{ GeV} \tag{1.25}$$

Besides the terms written above, the integration Lagrangian also describes the self-interaction of the scalar mesons σ and their interaction with leptons. Since the mass of the σ-meson is a free parameter, it may be chosen to be so large that at attainable energies processes involving the σ meson are strongly suppressed. This mass, however, cannot be considered to be arbitrarily large: in the limit $m_\sigma \to \infty$, the amplitudes to which diagrams with internal σ-lines correspond tend to infinity.

The muon part of the interaction Lagrangian has a form absolutely identical to (1.20).

The most interesting experimental prediction of the Weinberg–Salam model is the existence of neutral currents. This prediction has been brilliantly confirmed.

As concerns the interaction of charged weak currents, the predictions of the Weinberg–Salam model in lowest order coincide with the predictions of the phenomenological four-fermion model. But unlike the latter, the Weinberg–Salam model allows one also to calculate the radiative corrections of higher orders.

Let us discuss in more detail the renormalization of the Weinberg–Salam model. Since the Lagrangian (1.6), (1.11) is gauge-invariant, the renormalization procedure described in the preceding chapter can be applied to it. However, the gauge group contains the Abelian subgroup $U(1)$ and, in accordance with the classification of Section 4.8, is anomalous. Therefore, although formally gauge-invariant, the Weinberg–Salam model described by the Lagrangian (1.6), (1.11) is not renormalizable. The situation may be improved by means of the mechanism described in the preceding chapter. As we saw, anomalies are absent in the case of any gauge group, if the right-hand- and the left-hand-polarized fermions give to the anomalous triangle diagram contributions of equal magnitudes and opposite signs. Therefore, the introduction in the Weinberg–Salam model, in addition to the electron multiplets (1.1), of multiplets with opposite helicities

$$\tilde{R} = \frac{1}{2}(1-\gamma_5)\begin{pmatrix} N \\ E \end{pmatrix}, \quad \tilde{L} = \frac{1}{2}(1+\gamma_5)E, \tag{1.26}$$

interacting with vector fields in the same way as L and R, leads to the absence of anomalies in such a modified model. Note, however, that for the $\tilde{L}$ and $\tilde{R}$ it is not possible to use the muon and muon neutrino, since the "compensating" leptons must be involved in the weak interaction with opposite helicities. Thus, a renormalized extension of the Weinberg–Salam model requires the introduction of heavy leptons.

The second possibility, to be discussed further, is based on the use of mutual compensation of the anomalies of the lepton and hadron currents.

Let us now pass to the discussion of the weak and electromagnetic interactions of hadrons. Similarly to the Weinberg–Salam model, which predicts the existence of neutral lepton currents, the analogous model for the hadron sector predicts the existence of weak neutral hadron currents. If the hadron symmetry group is the group $SU(3)$, then the neutral current contains strangeness-changing terms. To verify this, we recall that in the $SU(3)$-symmetric theory the weak charged hadron current is described by the Cabbibo formula

$$j_\mu^+ = \bar{p}\gamma_\mu (1 + \gamma_5)(n \cos\theta + \lambda \sin\theta). \tag{1.27}$$

Here p and n are the proton and neutron quarks with charges ⅔ and −⅓, λ is the strange quark with a charge −⅓, and θ is the Cabbibo angle, characterizing the relative probabilities of processes with and without a change of strangeness.

As in the Weinberg–Salam model for leptons, the generators corresponding to the charged currents generate the group $SU(2)$. Therefore the gauge-invariant theory involves, together with the charged currents (1.27), a neutral current of the form

$$\begin{aligned} j_\mu^3 = {} & \bar{p}\gamma_\mu (1 + \gamma_5) p + \\ & + (\bar{n}\cos\theta + \bar{\lambda}\sin\theta)\gamma_\mu (1 + \gamma_5)(n\cos\theta + \lambda\sin\theta). \end{aligned} \tag{1.28}$$

The current j_μ^3 interacts with the third component of the Yang–Mills field A_μ^3 and consequently represents a linear combination of the electromagnetic and weak charged currents. As a result such a model allows processes involving neutral strangeness-changing currents, such as

$$K_L^0 \to \mu^+\mu^-, \quad K^+ \to \pi^+\nu\bar{\nu}, \tag{1.29}$$

and the probabilities of these processes should be comparable to the probabilities of processes involving charged currents. From experiments it is known that processes such as (1.29) are forbidden to a very high degree of accuracy. The ratio of the decay probability for $K_L^0 \to \mu^+\mu^-$ to the probability of the decay $K^+ \to \mu^+\nu_\mu$, involving charged currents, is $\leq 10^{-9}$. It is possible to forbid such processes in the gauge-invariant theory by giving up the assumption of the $SU(3)$ structure of hadrons. The simplest possibility is to substitute the group $SU(4)$ for the group $SU(3)$. In the quark model this is equivalent to the introduction of the fourth quark p' with a new quantum number—the "charm".

The four-quark gauge model of weak and electromagnetic interactions is constructed in the same way as the Weinberg–Salam model for leptons. The left-hand-polarized quarks are united in two $SU(2)$ doublets

$$L_1 = \frac{1}{2}(1+\gamma_5)\begin{pmatrix} p \\ n\cos\theta + \lambda\sin\theta \end{pmatrix};$$
$$L_2 = \frac{1}{2}(1+\gamma_5)\begin{pmatrix} p' \\ -n\sin\theta + \lambda\cos\theta \end{pmatrix}, \qquad 1.30)$$

and the right-hand-polarized ones in the singlets

$$R_1 = \frac{1}{2}(1-\gamma_5)p, \quad R_2 = \frac{1}{2}(1-\gamma_5)p',$$
$$R_3 = \frac{1}{2}(1-\gamma_5)(n\cos\theta + \lambda\sin\theta); \qquad (1.31)$$
$$R_4 = \frac{1}{2}(1-\gamma_5)(-n\sin\theta + \lambda\cos\theta).$$

The charged hadron current has the form

$$j_\mu^+ =$$
$$= \bar{p}_L\gamma_\mu(n\cos\theta + \lambda\sin\theta) + \bar{p}'_L\gamma_\mu(-n\sin\theta + \lambda\cos\theta) \qquad (1.32)$$

Commuting j_μ^+ and j_μ^-, we obtain for j_μ^3 the expression

$$j_\mu^3(x) = \left[\int j_0^+(y)\,d^3y,\ j_\mu^-(x)\right] =$$
$$= \bar{p}\gamma_\mu(1+\gamma_5)p + \bar{n}\gamma_\mu(1+\gamma_5)n + \bar{\lambda}\gamma_\mu(1+\gamma_5)\lambda. \quad (1.33)$$

In this current strangeness-changing terms are absent, and consequently processes such as (1.29) are forbidden in the lowest order of the weak interaction.

We shall not write out here the total gauge-invariant Lagrangian for the weak and electromagnetic interactions of hadrons. It is entirely analogous to the Lagrangian (1.3). Its most remarkable feature is the prediction of "charmed" hadron states. Recent experiments have confirmed this prediction of the gauge theories also.

To complete the description of the models of weak hadron interactions we recall that, according to the widely accepted point of view, there exist three varieties of quarks, differing from one another by their "color", that is each quark p, p', n, λ can assume three different "colors". Weak interactions are not sensitive to the color, and the corresponding Lagrangian is a sum of three identical Lagrangians. The hypothesis of the existence of color was put forward in order to explain the observed spectrum of hadrons within the framework of the usual assumptions on the relationship between the spin and statistics. Remarkably, it turned out that introduction of color at the same time makes the unified model of weak interactions, described above, self-consistent. In the model involving four leptons μ, e, ν_μ, ν_e and four quarks of three colors p, p', n, λ with charges $2/3$, $2/3$, $-1/3$, $-1/3$, anomalies are absent, and consequently the corresponding theory is renormalizable. In this case the total lepton charge (-2) is equal in magnitude and opposite in sign to the total charge of the quarks ($2/3 \times 3 = 2$), and for this reason the anomalies of the lepton and hadron currents are compensated.

The model described, until recently, has explained all known experimental facts. From a theoretical point of view it is distinguished in that it (supposing that quarks have fractional charges) is the only renormalizable model of weak interactions involving only the light leptons μ, e, ν_μ, ν_e. However, it is now known that the spectrum of leptons is not limited to the muon and electron and there exist also "heavy" leptons. Therefore a model with four leptons and four quarks of three colors is insufficient. There exist many possibilities for the construction of renormalizable gauge models, involving a large number of leptons and quarks. Since at present there is no reliable experimental information allowing one to give preference to any one of the concrete models, we shall not discuss them here.

5.2. ASYMPTOTIC FREEDOM. GAUGE THEORIES OF STRONG INTERACTIONS

At first sight the dynamics of strong inteactions seems to be too complicated to try to describe it in the framework of any reasonable quantum-field-theory model. Until recently, for the description of strong interactions either dispersion-relation methods, based on the most general physical requirements of causality and unitarity, or phenomenological models have been used. Attempts to construct a relativistic Lagrange model, which would give a detailed description of the dynamics of strong interactions, have not led even to qualitative results.

On the other hand, deep-inelastic lepton–proton scattering experiments yield evidence that a simple dynamical mechanism forms the basis of strong interactions. At large momentum transfers, which are equivalent to small spatial distances, hadrons behave as if they consisted of noninteracting point objects. Thus, the following qualitative picture arises: hadrons are composite objects, and the interaction between their components tends to zero at small distances. At the same time, at large distances the effective interaction becomes strong, so that a hadron is a strongly bound system.

It is possible to describe such an interaction in the framework of any quantum-field-theory model? The answer to this question turns out to be unambiguous. The above-described behavior of the interaction can be obtained only in a non-Abelian gauge theory. All consistent field-theory models which do not involve the Yang–Mills field lead to an increase of the effective interaction at small distances. This unique feature of the Yang–Mills fields is due to the phenomenon of asymptotic freedom, to the description of which we shall now pass.

We shall now discuss the asymptotic behavior of the Green functions in the deep-Euclidean region where the squares of all the momentum arguments p_i are negative and have large absolute values. This asymptotic behavior has no direct physical meaning, of course, since for the calculation of the S-matrix the values of the Green functions at $p_i^2 = m_i^2 \geq 0$ are needed. However, it may be shown that the probabilities of deep-inelastic scattering processes are directly related to the behavior of the Green functions in the deep-Euclidean region.

To be more precise, we shall investigate the asymptotic behavior of strongly connected proper vertex functions $\Gamma_m(\varkappa p_1, \ldots, \varkappa p_n, m, g)$, where $p_i^2 = -a_i^2 < 0$ as $\kappa \to \infty$. For this we shall need the technique of the renormalization group, the main concepts of which we shall briefly recall.

As we already know, the subtraction of the leading terms of the Taylor expansion of divergent proper vertex functions is equivalent to the insertion in the Lagrangian of local conterterms, which in turn is equivalent to the renormalization of the parameters involved in the Lagrangian. The transition from one subtraction point to another is equivalent to a finite renormalization. For instance, the insertion of the counterterms

$$\frac{1}{8}\operatorname{tr}\{(z_2-1)(\partial_\nu\mathcal{A}_\mu-\partial_\mu\mathcal{A}_\nu)^2+ \\ +2g(z_1-1)(\partial_\nu\mathcal{A}_\mu-\partial_\mu\mathcal{A}_\nu)[\mathcal{A}_\mu,\mathcal{A}_\nu]+ \\ +(z_1^2z_2^{-1}-1)[\mathcal{A}_\mu,\mathcal{A}_\nu]^2\}+\ \dots \quad (2.1)$$

(where $\cdots$ stands for the corresponding counterterms for fictitious particles and the fields of matter) is equivalent to the following renormalizations of the Green functions and the charges:

$$\begin{aligned} G^{\rm tr}_{\mu\nu}(k,g) &\to z_2^{-1}G^{\rm tr}_{\mu\nu}(k,g'), \\ \Gamma_{A^3}(p,q,g) &\to z_1\Gamma_{A^3}(p,q,g'), \\ \Gamma_{A^4}(k,p,q,g) &\to z_1^2z_2^{-1}\Gamma_{A^4}(k,p,q,g'), \\ g \to g' &= z_1z_2^{-3/2}g. \end{aligned} \qquad (2.2)$$

Therefore, the simultaneous insertion of counterterms (2.1) and multiplication of the coupling constant by $z_1^{-1}z_2^{3/2}$ does not change the renormalized coupling constant.

We shall denote the scalar functions appearing after the tensor structures in $G^{\rm tr}_{\mu\nu}$, Γ_{A^3} and Γ_{A^4} have been singled out as D, $g\Gamma_3$, $g^2\Gamma_4$. These functions are dimensionless and therefore may be represented in the form

$$D=D\left(\frac{k^2}{\lambda},\frac{m^2}{\lambda},g\right);\quad \Gamma_3=\Gamma_3\left(\frac{k_1^2}{\lambda}\dots\frac{k_3^2}{\lambda},\frac{m^2}{\lambda},g\right);$$
$$\Gamma_4=\Gamma_4\left(\frac{k_4^2}{\lambda}\dots\frac{k_{10}^2}{\lambda},\frac{m^2}{\lambda},g\right),$$
$$k_1^2\equiv p^2,\quad k_2^2\equiv q^2,\quad k_3^2\equiv(p+q)^2\quad \text{etc.,} \qquad (2.3)$$

where λ is the subtraction point. (The invariant variables are chosen so that the functions Γ_i are real at $k_i^2=\lambda<0$.) Then the condition that the theory is independent of the choice of the subtraction point, if a change in the subtraction point is accompanied by a simultaneous compensating charge transformation (the renormalization invariance), can be written in the form

$$D\left(\frac{k^2}{\lambda_2}, \frac{m^2}{\lambda_2}, g_2\right) = z_2 D\left(\frac{k^2}{\lambda_1}, \frac{m^2}{\lambda_1}, g_1\right),$$

$$\Gamma_3\left(\frac{k_1^2}{\lambda_2} \dots \frac{k_3^2}{\lambda_2}, \frac{m^2}{\lambda_2}, g_2\right) =$$

$$= z_1^{-1}\Gamma_3\left(\frac{k_1^2}{\lambda_1} \dots \frac{k_3^2}{\lambda_1}, \frac{m^2}{\lambda_1}, g_1\right), \quad (2.4)$$

$$\Gamma_4\left(\frac{k_4^2}{\lambda_2} \dots \frac{k_{10}^2}{\lambda_2}, \frac{m^2}{\lambda_2}, g_2\right) =$$

$$= z_1^{-2} z_2^{!}\Gamma_4\left(\frac{k_4^2}{\lambda_1} \dots \frac{k_{10}^2}{\lambda_1}, \frac{m^2}{\lambda_1}, g_1\right),$$

$$g_2 = z_1 z_2^{-1/2} g_1.$$

We shall consider these functions to be normalized by the condition

$$D, \ \Gamma_3, \ \Gamma_4 = 1 \quad (2.5)$$

at $x_i = k_i^2/\lambda = 1$.

From the equations (2.4) and the normalization condition it follows that

$$z_2 = D\left(\frac{\lambda_1}{\lambda_2}, \frac{m^2}{\lambda_2}, g_2\right),$$

$$z_1^{-1} = \Gamma_3\left(\frac{\lambda_1}{\lambda_2}, \frac{\lambda_1}{\lambda_2}, \frac{\lambda_1}{\lambda_2}, \frac{m^2}{\lambda_2}, g_2\right). \quad (2.6)$$

Therefore, by introducing the dimensionless variables

$$x_i = \frac{k_i^2}{\lambda_2}, \quad y = \frac{m^2}{\lambda_2}, \quad t = \frac{\lambda_1}{\lambda_2}, \quad (2.7)$$

the system (2.4) can be written as

$$D(x, y, g) = D(t, y, g)\, D\left(\frac{x}{t}, \frac{y}{t}, \bar{g}(t, y, g)\right),$$

$$\Gamma_3(x_1 \dots x_3, y, g) =$$

$$= \Gamma_3(t \dots t, y, g)\, \Gamma_3\left(\frac{x_1}{t} \dots \frac{x_3}{t}, \frac{y}{t}, \bar{g}(t, y, g)\right),$$

$$\Gamma_4(x_4 \dots x_{10}, y, g) =$$

$$= \Gamma_4(t \dots t, y, g)\, \Gamma_4\left(\frac{x_4}{t} \dots \frac{x_{10}}{t}, \frac{y}{t}, \bar{g}(t, y, g)\right). \quad (2.8)$$

where the function $\bar{g}(t, y, g)$

$$\bar{g}(t, y, g) = g\Gamma_3(t \ldots t, y, g)[D(t, y, g)]^{3/2} \tag{2.9}$$

is invariant under the transformations (2.4). This function is called the invariant charge.

We shall deal with these equations in the deep-Euclidean region $x_i = \varkappa\tilde{x}_i$, $\varkappa \to \infty$. We shall also assume that

$$|\lambda_i| \gg m^2. \tag{2.10}$$

It can be shown that in the renormalizable theories the main terms in the asymptotic behavior of the Green functions in the indicated region are independent of the mass, and therefore in the equations (2.8) one can put $y = 0$.

It is convenient to rewrite the system (2.8) in the differentiatial form. Differentiating (2.8) with respect to t and assuming $t = 1$, we obtain

$$\begin{aligned}
&\left(\varkappa\frac{\partial}{\partial\varkappa} - \beta(g)\frac{\partial}{\partial g}\right) D(\varkappa\tilde{x}_i, g) = \psi_2(g) D(\varkappa\tilde{x}_i, g),\\
&\left(\varkappa\frac{\partial}{\partial\varkappa} - \beta(g)\frac{\partial}{\partial g}\right) \Gamma_3(\varkappa\tilde{x}_i, g) = \psi_3(g) \Gamma_3(\varkappa\tilde{x}_i, g),\\
&\left(\varkappa\frac{\partial}{\partial\varkappa} - \beta(g)\frac{\partial}{\partial g}\right) \Gamma_4(\varkappa\tilde{x}_i, g) = \psi_4(g) \Gamma_4(\varkappa\tilde{x}_i, g),
\end{aligned} \tag{2.11}$$

where

$$\beta(g) = \left.\frac{\partial\bar{g}(t, g)}{\partial t}\right|_{t=1}, \tag{2.12}$$

$$\psi_i(g) = \left.\frac{\partial\Gamma_i(t \ldots t, g)}{\partial t}\right|_{t=1}; \quad \psi_2(g) = \left.\frac{\partial D(t, g)}{\partial t}\right|_{t=1}. \tag{2.13}$$

Analogous equations may obviously be written also for the higher-order Green functions $\Gamma_n(\varkappa\tilde{x}_1, \ldots, \varkappa\tilde{x}_l, g)$. The invariant charge $\bar{g}$ satisfies the simplest equation. Differentiating the invariance condition

$$\bar{g}(\varkappa, g) = \bar{g}\left(\frac{\varkappa}{t}, \bar{g}(t, g)\right) \tag{2.14}$$

with respect to t and assuming $t = 1$, we have

$$\left(\varkappa\frac{\partial}{\partial\varkappa} - \beta(g)\frac{\partial}{\partial g}\right)\bar{g}(\varkappa, g) = 0. \tag{2.15}$$

The boundary condition has the form

$$\bar{g}(1, g) = g. \tag{2.16}$$

It is possible to obtain another useful form of the equation for the invariant charge by differentiating (2.14) with respect to $\varkappa$ and then assuming $\varkappa = t$. Thus, we obtain

$$\varkappa \frac{\partial \bar{g}}{\partial \varkappa} = \beta(\bar{g}) \tag{2.17}$$

or in the integral form

$$\int\limits_{g}^{\bar{g}(\varkappa, g)} \frac{da}{\beta(a)} = \ln \varkappa. \tag{2.18}$$

By means of the equation (2.15) it is easy to express the solution of the system (2.11) in terms of the invariant charge. The general solution has the form

$$\Gamma_n(\varkappa \tilde{x}_i, g) = \Gamma_n(\tilde{x}_i, \bar{g}) \exp\left\{ \int\limits_1^{\varkappa} d\varkappa' \psi_n[\bar{g}(\varkappa', g)] (\varkappa')^{-1} \right\}. \tag{2.19}$$

From this formula an important conclusion follows: in the asymptotic region the invariant charge $\bar{g}$ is the effective parameter characterizing the interaction strength. Therefore, to obtain information about the asymptotic behavior of the Green function, it is necessary to know the behavior of $\bar{g}$. From the equations (2.15) and (2.17) it follows that the behavior of the invariant charge is defined by the properties of the function $\beta(\bar{g})$. If $\beta(\bar{g})$ is positive, then the invariant charge increases with $\varkappa$. If at some value of $\bar{g}$ the function $\beta(\bar{g})$ turns to zero and the integral on the left-hand side of (2.18) diverges, the the right-hand side tends to infinity. In other words, as $\varkappa \to \infty$, $\bar{g}(\varkappa, g) \to \bar{g}_0$, where $\bar{g}_0$ is the zero of the function β. If the function β has no zeros at $\bar{g} > g$, then as $\varkappa \to \infty$, $\bar{g} \to \infty$. For a negative $\beta(\bar{g})$ the situation is reversed. The function $\bar{g}(\varkappa, g)$ decreases as the κ increases. If the function $\beta(\bar{g})$ turns to zero at $\bar{g} = \bar{g}_0 < g$, then as $\varkappa \to \infty$, $\bar{g} \to \bar{g}_0$.

Thus, the zeros of the function β can be stable and unstable. If the constant g is in the vicinity of a "stable" zero $\bar{g}_0$, then as $\varkappa$ increases the invariant charge $\bar{g}(\varkappa, g)$ tends to $\bar{g}_0$. In the unstable situation, as κ increases the invariant charge departs farther and farther from $\bar{g}_0$. It tends either to the next zero or to infinity. Both these cases are presented graphically in Figure 19. From this figure it may be seen that the stable and unstable zeros alternate with each other.

In practice the only reliable way of calculating the β-function is by perturbation theory. Therefore, in reality we can judge its behavior only in the vicinity of the point $g = 0$. If in the vicinity of $g = 0$ the β function is negative, then as $\varkappa$ increases the invariant charge tends to zero. In this case it is said that the zero is an ultraviolet-stable point, and the theory is asymptotically free. This last statement signifies that as the energy increases the effective interaction becomes weaker and weaker, and at small distances the particles behave as free ones. In the case, when the β-function is positive near zero, the effective charge increases with the energy, as a result of which we go beyond the range of applicability of perturbation theory.

In most quantum-field-theory models a second possibility is realized. For example, in electrodynamics in the lowest order of α

$$\beta(\alpha) = \frac{\alpha^2}{3\pi}. \tag{2.20}$$

Substituting this value into the formula (2.18), we obtain

$$\bar{\alpha}(\varkappa, \alpha) = \frac{\alpha}{1 - \frac{\alpha}{3\pi} \ln \varkappa}. \tag{2.21}$$

As is seen, as $\varkappa$ increases $\bar{\alpha}(\varkappa, \alpha)$ increases, and at $\varkappa = e^{3\pi/\alpha}$ it goes to infinity. Of course, in reality at $\varkappa \sim e^{3\pi/\alpha}$ the formula (2.21) cannot be used, since the function β has been calculated assuming the effective coupling constant to be small.

If we nevertheless try to extrapolate the formula (2.21) to the region of large $\bar{\alpha}$, we immediately come to a contradiction. In the electrodynamics the invariant charge is related to the photon Green

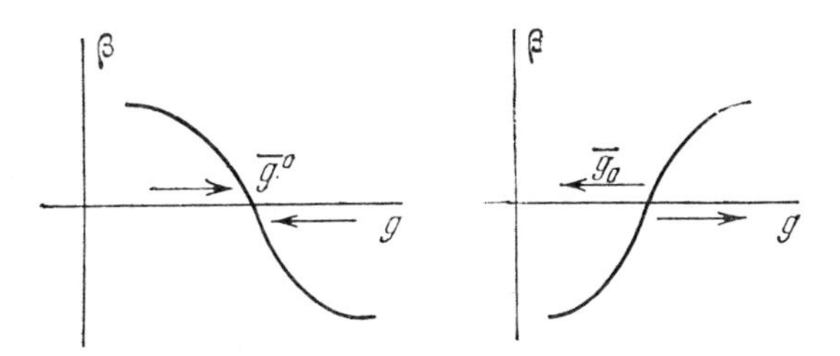

Fig. 19. Stable and unstable zeros of the β-function.

function by

$$\bar{\alpha}(\varkappa, \alpha) = \alpha d(\varkappa, \alpha), \tag{2.22}$$

where

$$D_{\mu\nu}(k) = -\frac{1}{k^2}\left(g_{\mu\nu} - \frac{k_\mu k_\nu}{k^2}\right) d(k^2, \alpha). \tag{2.23}$$

Therefore, if the denominator in the expression (2.21) becomes equal to zero, that means the photon Green function has a pole. It is not difficult to verify that the residue at this pole is negative. The corresponding state has a negative norm, which is incompatible with the unitarity condition. Thus, in the case of $\beta(g) > 0$ at $g \sim 0$, the perturbation theory cannot give any reliable information about the asymptotic behavior of the Green functions.

Quite different is the case of the Yang–Mills theory. In this theory $\beta(g)$ is negative in the vicinity of zero, and consequently zero is an ultraviolet-stable point. Indeed, by definition

$$\beta(g) = \left.\frac{\partial \bar{g}^2(\varkappa, g)}{\partial \varkappa}\right|_{\varkappa=1}, \tag{2.24}$$

where in the case of the Yang–Mills field the invariant charge is equal to

$$\bar{g}^2(\varkappa, g) = g^2 \Gamma_3^2 D^3(\varkappa). \tag{2.25}$$

Since $\varkappa = k^2/\lambda$,

$$\left.\frac{\partial}{\partial \varkappa}\right|_{\varkappa=1} = -\lambda \left.\frac{\partial}{\partial \lambda}\right|_{\lambda=k^2} = -\left.\frac{\partial}{\partial \ln \lambda}\right|_{\lambda=k^2}. \tag{2.26}$$

On the other hand,

$$\frac{\partial}{\partial \ln \lambda} D\left(\frac{k^2}{\lambda}\right)_{\lambda=k^2} = -\frac{\partial}{\partial \ln \lambda} z_2^{-1}\left(\frac{\Lambda}{\lambda}\right)\Big|_{\lambda=\Lambda}, \tag{2.27}$$

$$\frac{\partial}{\partial \ln \lambda} \Gamma_3\left(\frac{k^2}{\lambda}\right)\Big|_{\lambda=k^2} = -\frac{\partial}{\partial \ln \lambda} z_1\left(\frac{\Lambda}{\lambda}\right)\Big|_{\lambda=\Lambda}. \tag{2.28}$$

Therefore, for the definition of $\beta(g)$ we can use the values of z_i determined previously. Thus we obtain

$$\beta(g^2) = -\frac{22}{3}\frac{g^4}{(4\pi)^2}. \tag{2.29}$$

Consequently, the square of the invariant charge tends to zero as $\varkappa \to \infty$:

$$\bar{g}^2(\varkappa, g^2) = \frac{1}{1 + \frac{g^4}{(4\pi)^2} \frac{22}{3} \ln \varkappa}. \tag{2.30}$$

In the deep-Euclidean region the interaction "dies out" and the theory behaves as free. For the case of an arbitrary gauge group, and taking into account the interaction with the fields of matter, the function $\beta(g)$ is given by the formula

$$\beta(g^2) = \left[-\frac{11}{3} C(G) + \frac{4}{3} T(R)\right] \frac{g^2}{16\pi^2}, \tag{2.31}$$

where

$$\delta^{ab} C(G) = t^{acd} t^{bcd}; \quad T(R)\, \delta^{ab} = \mathrm{tr}\, \{\Gamma^a, \Gamma^b\}, \tag{2.32}$$

where t^{acd} are the structure constants of the group and Γ^i are the generators of the representation realized by the fields of matter. If the number of multiplets of the fields of matter is not too large, then in this case also the theory is asymptotically free.

Thus, if the Yang–Mills fields are the carriers of the strong interactions, then at small distances quasifree particles will indeed be observed, in agreement with deep-inelastic scattering experiments.

Contrariwise, at $\varkappa < 1$ the effective coupling constant increases. Of course, in this case the formula (2.30) obtained by means of the perturbation theory cannot be used. Nevertheless, if the β-function has no zeros at $g > 0$, the it follows from the equation (2.18) that

$$\bar{g}(\varkappa, g) \to \infty, \quad \varkappa \to 0. \tag{2.33}$$

Such a behavior of the invariant charge would mean that with increasing the distance the interaction strength would increase indefinitely, and consequently the particles would not be able to withdraw from each other to large distances.

The qualitative picture described above is realized in the hypothetical model of strong interactions known as "quantum chromodynamics". In this model hadrons are considered to be quark bound states. There exist several types of quarks, differing from each other by the quantum number "flavor". Strangeness and charm are examples of "flavors". Each quark in its turn can exist in three varieties, differing in "color". Thus, the quarks are represented by the following matrix:

$$\begin{Bmatrix} u_r, & u_y, & u_b \\ d_r, & d_y, & d_b \\ c_r, & c_y, & c_b \\ s_r, & s_y, & s_b \\ \cdot & \cdot \cdot \cdot & \cdot \end{Bmatrix}. \tag{2.34}$$

Here the indices r, y, b stand for the "colors" (red, yellow, blue) and the letters u, d, c, s indicate the various flavors. The interaction between quarks is due to the exchange of colored Yang–Mills fields, that is, "gluons". The gauge group SU_3 acts in the color space. The Yang–Mills fields form a color octet and are neutral with respect to "flavors". The strong-interaction Lagrangian has the form

$$\mathscr{L} = \frac{1}{8}\operatorname{tr}\{\mathscr{F}_{\mu\nu}\mathscr{F}_{\mu\nu}\} + \bar{q}\,\{i\gamma_\mu\,[\partial_\mu - g\Gamma\,(\mathscr{A}_\mu)] - m\}\,q;$$
$$q = u,\ d,\ \ldots \tag{2.35}$$

The color SU_3^c symmetry is assumed to be exact. This means that the Yang–Mills fields have a mass exactly equal to zero.

The observed spectrum of hadrons is generated by colorless quark bound states, which are singlets with respect to the group SU_3^c. In the approximation where all quarks have equal masses, the Lagrangian (2.35) is invariant under the transformations of the group SU^f, acting in the space of flavors. Therefore, it is convenient to classify the spectrum of baryons with respect to the group SU^f [until recently the most popular candidate for the SU^f group was the group $SU(4)$]. In reality the symmetry of the SU^f is broken, so degeneracy in the masses within the hadron multiplets is absent. The most problematic thing in this scheme is why quarks are not observed experimentally and why, notwithstanding the fact that the Yang–Mills fields have zero mass, the strong interactions have a finite range. For an explanation the quark-confinement hypothesis, based on the assumption that the theory described by the Lagrangian (2.35) is asymptotically free, has been put forward. In the spirit of the discussion following Equation (2.10) it is assumed that due to the asymptotic freedom the effective coupling constant increases infinitely as the distance between the interacting objects increases. As a result, the colored objects—quarks and gluons—can never withdraw from each other to macroscopic distances. Only colorless bound states corresponding to real hadrons are observable. The effective interaction of these bound complexes has a finite range and it is just this range which is observed in experiments at moderate energies.

BIBLIOGRAPHY

NOTES

As was pointed out in the preface, our book is supplementary to the already existing manuals on quantum field theory, the closest to ours being the monograph "An Introduction to the Theory of Quantized Fields" [1] by N. N. Bogolubov and D. V. Shirkov. Unlike most handbooks on quantum field theory, this one describes the quantum dynamics mainly by means of the path integral. The application of this method to quantum-mechanical problems is expounded in the book of R. Feynman and A. Hibbs [2], and the monographs recently published by A. N. Vasil'ev [3] and V. N. Popov [4] are dedicated to the use of this method in the theory of systems with an infinite number of degrees of freedom. The classical geometrical aspects of gauge fields have been considered in a monograph by N. P. Konoplyova and V. N. Popov [5], and their quantization and application to elementary-particle models have been briefly described in the book by J. Taylor [6].

Chapter I

Gauge fields were first introduced in physics in the work of C. N. Yang and R. L. Mills [7] for fields carrying the interaction of isotopic spins. The natural generalization to the case of internal degrees of freedom of a more general nature is discussed, for example, in the publications [8,9,10].

Feynman was the first to draw attention to the specific character of the quantization of non-Abelian gauge fields [11]. His approach, based on the reconstruction of diagrams with loops by means of tree diagrams, was developed by De Witt [12], who formulated the final rules for quantization of gauge fields and gravity fields in [13]. An independent derivation of the rules of perturbation theory for these theories, based on the path integration, was obtained by L. D. Faddeev and V. N.Popov in [14] (see also [15]). The publications [16,17,18] are also dedicated to the construction of the perturbation theory for gauge fields. The hypothesis stated in Feynman's lectures [11] that the perturbation theory for gauge fields may be obtained in the limit $m \to 0$ of the theory of massive vector fields turned out to be incorrect [19,20]. The first realistic unified interaction models, based on the Higgs mechanism [21,22,23], were formulated by S. Weinberg [24] and A. Salam [25].

In 1971 't Hooft extended the quantization procedure for Yang–Mills fields to the case of theories with spontaneously broken symmetry [26]. In 1971–1972, in a series of publications by A. A. Slavnov [27,28], J. Taylor [29], B. Lee and J. Zinn-Justin [30], and G. t'Hooft and M. Veltman [31], the methods were developed of invariant regularization and renormalization for the theory of gauge fields (including models with spontaneously broken symmetry), and thus the construction of the quantum theory of gauge fields in the framework of perturbation theory was completed. Various aspects of the theory of gauge fields and their applications have been presented in reports at international conferences on high-energy physics by B. Lee [32], J. Illiopoulos [33], and A. Slavnov [34].

From the viewpoint of differential geometry the classical Yang–Mills field represents a connection in the principal fiber-bundle space, the basis of which is the space-time manifold and a typical fiber of which is the internal symmetry group. The concept of connection, generalizing the Euclidean connection in the Riemann space, has been developed starting from the twenties in the work of many geometers, specifically, of H. Weyl and E. Cartan. Its modern formulation first appeared in the work of Ehresmann [35]. An excellent introduction to the theory of fiber bundles and connections may be found in the book by Lischnerovitch [36].

In the twenties, due to the success of the theory of general relativity, many attempts were made to geometrize the electromagnetic field. A correct view of this field as a part of the connection involved in the covariant derivative of the complex fields appeared in the work of H. Weyl [37] and V. A. Fock [38] on the formulation of the Dirac equation in the gravitational field. H. Weyl speaks directly about electrodynamics as the general relativity theory in the charge space.

The classical solutions of the equations of motion involving gauge fields have been a subject of intensive research during the last three years. We give references to some of the main publications in this field [39,40,41,42], in which solutions for the vacuum and the soliton sector are studied.

Chapter II

The path integral was first introduced for the formulation of quantum mechanics by Feynman. The history and main concepts may be found in the monograph [2]. The Feynman diagrams in the perturbation theory, first introduced in [43], were substantiated by means of the path integral in [44]. The monographs [3,4] contain a more up-to-date review of the applications of the path-integral method in quantum physics. The interpretation of the path-integral method in the present book follows the lectures of one of the authors [45]. The introduction of the path integral in quantum mechanics by the formula (2.1.12) is adopted from the work of Tobocman [46]. The holomorphic representation of quantum mechanics is from the work of V. A. Fock; it appeared under the name of coherent states in quantum optics. Its mathematical formulation may be found in the monograph of F. A. Berezin [47]. There also the first rigorous exposition of the integration over anticommuting variables is given.

The boundary conditions in the path integral were considered by O. I. Zav'yalov [48] and A. N. Vasil'ev [3].

The Green functions were introduced in the quantum field theory by J. Schwinger [49]. The idea of reducing the calculation of the S-matrix to the problem of the calculation of the S-matrix for scattering by an external source also belongs to Schwinger [50].

The introduction of the path integral in terms of the Gaussian functional was outlined in the first edition of the monograph [1]. The exposition in the present book of the axiomatics of the path integral by means of the Gaussian functional follows the work of one of the authors [51]. A similar approach was developed also in [52].

Chapter III

The generalized Hamiltonian dynamics was first introduced by Dirac [53] (see also his lectures [54]). The Hamiltonian formulation of gauge theories in the Coulomb gauge was investigated by J. Schwinger [55]. The general formulation of the path integral in the generalized Hamiltonian form was given by one of the authors [56]. The Hamilton gauge $A_0 = 0$, less popular than the Coulomb one $\partial_i A_i = 0$, formed the basis for the construction of the gauge field theory in Feynman's lectures [57].

The change-of-variables method in the path integral to pass from one gauge to another was proposed in [14]. Its geometrical interpretation in terms of various parametrizations of gauge-equivalent classes is discussed in [56,58]. The generalized α-gauges were first considered accurately in [13] (see also [17,56]). The method of transition to the α-gauge, described in the present book, is adopted from the work of G. 't Hooft [26].

Chapter IV

The renormalization theory goes back to the ideas of H. A. Kramers [59] and H. A. Bethe [60]. A number of authors, including R. P. Feynman, J. Schwinger, F. Dyson, A. Salam, and others, took part in its development. The complicated mathematically rigorous renormalization theory (the theory of R-operation) was first constructed in the work of N. N. Bogolubov and O. S. Parasyuk [61]. An excellent exposition of the R-operation may be found in the monograph [1], where a detailed bibliography on the renormalization theory is also presented.

Regularization by means of higher covariant derivatives was first propounded in the work of A. A. Slavnov [62] and then applied to the Yang–Mills theory in [28,30]. The additional regularization of one-loop diagrams, described in Section 4.4, was constructed in [63].

The dimensional regularization was proposed in the work by G. 't Hooft and M. Veltman [31], C. G. Bollini and J. G. Giabidgi [64], and J. F. Ashmore [65].

The identities relating two- and three-point Green functions in quantum electrodynamics were first obtained by J. C. Ward [66]. The generalized relations, connecting any Green function with the function containing one external photon line less, were obtained by E. S. Fradkin [67] and Y. Takahashi [68]. The electrodynamical Ward identities are not generalizable directly to the case of non-Abelian gauge fields. In the non-Abelian theory their role is played by the so-called generalized Ward identities, first obtained by A. A. Slavnov [27] and J. C. Taylor [29]. Their derivation, given in the present book, follows [69]. An alternative derivation based on the use of the invariance of the effective Lagrangian with respect to some transformation with anticommuting parameters (supertransformation) was proposed by C. Becchi, A. Rouet, and R. Stora [70]. In the literature the generalized Ward identities for single-particle irreducible Green functions, obtained by B. W. Lee [71] (see also [52]), are also used. The structure of the renormalized action was investigated in [27,29,30,31], as a result of which there appeared a proof of gauge-invariance and unitarity of the renormalized S-matrix. Another approach to the renormalization of gauge theories, based on the use of the formalism of the Zimmerman normal products, was developed by C. Becchi, A. Rouet, and R. Stora [70,72].

The dependence of the renormalization constants and the Green functions on the choice of the gauge condition is discussed in detail in the work by R. Callosh and I. Tyutin [73].

The anomalous Ward identities were first studied by S. L. Adler

[74] and by S. Bell and R. Jackiw [75]. Their role in the problem of the renormalization of gauge theories was discussed in [76–79]. The classification of anomalous interactions presented in this book follows the work by H. Georgi and S. Glashow [79].

Chapter V

The first (and until now the most popular) realistic unified model of weak and electromagnetic interactions was constructed by S. Weinberg [24] and A. Salam [25]. There exist a large number of unified models differing from the Weinberg–Salam model either in the choice of the gauge group or in the multiplets involved. Some of them are presented in review reports [32,33,34]. At present the choice of one or another of the models has not been settled.

The mechanism of suppression of the strangeness-changing neutral currents by the introduction of a new "charmed" quark was propounded by S. Glashow, J. Illiopoulos, and L. Maiani [80].

The hypothesis of the existence of an additional degree of freedom of the quarks, which became known as "color", was first proposed in the work of O. Greenberg [80], of N. N. Bogolubov, B. V. Struminsky, and A. N. Tavkhelidze [82], and of M. Y. Han and Y. Nambu [83] for the explanation of the hadron statistics. In the work of J. Pati and A. Salam [84], of H. Fritzsch, M. Gell-Mann, and H. Leutwyller [85], and S. Weinberg [86] it was first suggested that strong interactions take place owing to the exchange of Yang–Mills mesons interacting with the color degrees of freedom. The corresponding hypothetical model was named quantum chromodynamics

The group features of the renormalization transformations were first drawn attention to by E. C. G. Stueckelberg and A. Peterman [87]. The group of multiplicative renormalizations in quantum electrodynamics was used by M. Gell-Mann and F. Low [88] for the investigation of the ultraviolet asymptotics of the Green functions. The general theory of the renormalization group was constructed in the work of N. N. Bogolubov and D. V. Shirkov [89,90]. A detailed exposition of this theory may be found in the monograph [1]. The differential equations of the renormalization group were investigated by L. V. Ovsyannikov [91]. Analogous equations were obtained for the case of quantum field theory by C. Callan [92] and K. Symanzik [93]. The asymptotic freedom of the Yang–Mills fields was discovered by G. 't Hooft [94], by D. Gross and F. Wilczek [95], and by H. D. Politzer [96]. The hypothesis of "quark confinement" was discussed in [85,86,97,98].

REFERENCES

1. Bogolubov, N. N., Shirkov, D. V. *An Introduction to the Theory of Quantized Fields*, Moscow, Nauka, 1976.
2. Feynman, R. P., Hibbs, A. R., *Quantum Mechanics and Path Integrals*, New York, McGraw-Hill Book Company, 1965.
3. Vasil'ev, A. N., *Functional Methods in Quantum Field Theory and Statistics*, Leningrad, Leningrad University Press, 1976.
4. Popov, V. N., *Path Integrals in Quantum Field Theory and Statistical Physics*, Moscow, Atomizdat, 1976.
5. Konoplyova, N. P., Popov, V. N., *Gauge Fields*, Moscow, Atomizdat, 1972.
6. Taylor, J. C., *Gauge Theories of Weak Interactions*, Cambridge University Press, Cambridge, 1976.
7. Yang, C. N., Mills, R. L., Phys. Rev., **96**, 191, 1954.
8. Utiyama, R. Phys. Rev., **101**, 1597, 1956.
9. Salam, A., Ward, J. C., Nuovo Cimento, **XI**, 568 (1959).
10. Glashow, S. L., Gell-Mann, M., Ann. of Phys., **15**, 437 (1961).
11. Feynman, R., Acta Phys. Polonica, **24**, 697 (1963).
12. De Witt, B., Phys. Rev. Lett., **12**, 742, 1964.
13. De Witt, B., Phys. Rev., **160**, 1113; 1195 (1967).
14. Faddeev, L. D., Popov, V. N., Phys. Lett., **B25**, 30 (1967).
15. Faddeev, L. D., Popov, V. N., Preprint, Institute for Theoretical Physics, Ukrainian Acad. Sci., Kiev, 1967.
16. Mandelstam S., Phys. Rev., **175**, 1580 (1968).
17. Fradkin, E. S., Tyutin, I. V., Phys. Lett., **B30**, 562 (1969).
18. Vainstein, A. I., Khriplovich, I. B., Yadernaya Fizika, **13**, 198, 1971.
19. Boulware, A., Ann. Phys., **56**, 140, 1970.
20. Slavnov, A. A., Faddeev, L. D., TMP*, **3**, 18, 1970.
21. Higgs, P. W., Phys. Lett, **12**, 132, 1964.
22. Englert, F., Brout, R., Phys. Rev. Lett., **13**, 321, 1964.
23. Kibble, T. W. B., Phys. Rev., **155**, 1554, 1967.
24. Weinberg, S., Phys. Rev. Lett., **19**, 1264, 1967.
25. Salam, A., *Elementary Particle Theory*, N. Svartholm, (ed.) Stockholm. Almquist, Forlag AB, 1968.
26. Hooft, 't G. Nucl. Phys., **B35**, 167, 1971.
27. Slavnov, A. A., TMP, **10**, 99, 1972.
28. Slavnov, A. A., TMP, **13**, 174, 1972.
29. Taylor, J. C., Nucl. Phys., **B33**, 436, 1971.
30. Lee, B. W., Zinn-Justin, J., Phys. Rev. D, **5**, 3137, 1972.
31. Hooft't G. Veltman, M., Nucl. Phys., **B44**, 189; **B50**, 318, 1972.
32. Lee, B. W., Proceedings of the 14 International Conference on High Energy Physics, Batavia, 1972.
33. Illiopoulos, J., Proceedings of the 17 International Conference on High Energy Physics, London, 1974.

---

*TMP: Teoreticheskay'a i Matematicheskay'a Fizika.

34. Slavnov, A. A., Proceedings of the 18 International Conference on High Energy Physics, Tbilisi, 1976.
35. Ehresmann, Coll.top. Bruxelles, p. 29, 1950.
36. Lischnerovich, A., *Theory of Connections as a Whole and Holonomy Groups,* Moscow, Inostrannaya Literature, 1960 (Russian translation).
37. Weyl, H., Z. f. Phys., **56**, 330, 1929.
38. Fock, V., Journ. de Physique, **10**, 392, 1929.
39. Polyakov, A. M., Lett. to JETP, **20**, 430, 1974.
40. Hooft 't G. Nucl. Phys., **B79**, 2761, 1974.
41. Faddeev, L.D. Preprint MPI-RAE/Pth München, 1974.
42. Belavin, A. A., Polyakov, A. M., Tyupkin, Y., Schwarz, A. S. Phys. Lett. **B59**, 85, 1975.
43. Feynman, R. P., Rev. Mod. Phys., **20**, 367, 1948.
44. Feynman, R. P., Phys. Rev., **80**, 440, 1950.
45. Faddeev, L. D., Les Houches Lecture, Session 20. North-Holland, 1976.
46. Tobocman, W., Nuovo Cimento, **3**, 1213, 1956.
47. Berezin, F. A., *The Method of Secondary Quantization*, Moscow, Nauka, 1965.
48. Zav'yalov, O. I., Thesis, Math. Inst. Acad. Sci., Moscow, 1970.
49. Schwinger, J., Phys. Rev., **75**, 651, 1949.
50. Schwinger, J., Proc. Nat. Acad. Sci., **37**, 452, 1951.
51. Slavnov, A. A., TMP, **22**, 177, 1975.
52. Zinn-Justin, J., Lecture Notes in Physics, **37**, Berlin, Springer-Verlag,
53. Dirac, P. A. M., Proc. Roy. Sci., **A246**, 326, 1958.
54. Dirac, P. A. M., *Lectures on Quantum Mechanics*, New York, Yeshiva University, 1964.
55. Schwinger, J., Phys. Rev., **125**, 1043, 1962; **127**, 324, 1962.
56. Faddeev, L. D., TMP, **I**, 3, 1969.
57. Feynman, R., Les Houches Lectures, North-Holland, 1977.
58. Popov, V. N., Faddeev, L. D., Uspekhi Fiz. Nauk, **111**, 427, 1973.
59. Kramers, H. A., Rapports du 8^{e} Conseil Solvay, Bruxelles, 1950.
60. Bethe, H. A., Phys. Rev., **72**, 339, 1947.
61. Bogolubov, N. N., Parasyuk, O. S., Doklady Akad. Nauk USSR, **55**, 149, 1955; **100**, 429, 1955, Acta Math., **97**, 227, 1957.
62. Slavnov, A. A., TMP, **33**, 210, 1977.
64. Bollini, C. G., Giabidgi, J.T., Phys. Lett., **B40**, 566, 1972.
65. Ashmore, J. F., Nuovo Cimento Lett., **4**, 289, 1972.
66. Ward, J. C., Phys. Rev., **77**, 2931, 1950.
67. Fradkin, E. S., JETP, **29**, 288, 1955.
68. Takahashi, Y., Nuovo Cimento, **6**, 370, 1957.
69. Slavnov, A. A., Nucl. Phys., **B97**, 155, 1975.
70. Becchi, C., Rouet, A., Stora, R. Comm. Math. Phys., **42**, 127, 1975.
71. Lee, B. W., Phys. Lett., **B46**, 214, 1974; Phys. Rev., **9**, 933, 1974.
72. Becchi, C., Rouet, A., Stora, R. *Renormalization Theory* (G. Velo,; A. S. Wightman, eds.)—D. Reidel Publ. Co., 1976; Ann. Phys., **98**, 287, 1976.
73. Kallosh, R., Tyutin, I. Yadernaya Fizika, **17**, 190, 1973.
74. Adler, S. L. Phys. Rev., **177**, 2426, 1969.
75. Bell, S., Jackiw, R., Nuovo Cimento, **A60**, 47, 1969.
76. Slavnov, A. A., TMP, **7**, 13, 1971.
77. Gross, D. J., Jackiw, R., Phys. Rev. D, **6**, 477, 1972.

78. Bouchiat, C., Illiopoulos, J., Meyer, P. Phys. Lett., **B38**, 519, 1972.
79. Georgi H., Glashow S.L., Phys. Rev. D, **6**, 429, 1977.
80. Glashow, S. L., Illiopoulos, J., Maiani, L. Phys. Rev. D, **2**, 185, 1970.
81. Greenberg, O. W., Phys. Rev. Lett., **13**, 598, 1964.
82. Bogolubov, N. N., Struminsky, B. V., Tavkhelidze, A. N., JINR Preprint D-1968, 1965.
83. Han, M. Y., Nambu, Y., Phys. Rev. **139**, B1006, 1965.
84. Pati, J., Salam A., Phys. Rev. D, **8**, 1240, 1973.
85. Fritzsch, H., Gell-Mann, M., Leutwyller, H., Phys. Lett., **B47**, 365, 1973.
86. Weinberg, S., Phys. Rev. Lett., **31**, 494, 1973.
87. Stueckelberg, E. C. G., Peterman, A. Helv. Phys. Acta, **26**, 499, 1953.
88. Gell-Mann, M., Low F. Phys. Rev., **95**, 1300, 1954.
89. Bogolubov, N. N., Shirkov, D. V., Doklady Akad. Nauk SSSR, **103**, 203, 391, 1955.
90. Bogolubov, N. N., Shirkov, D. V., JETP, **30**, 77; Nuovo Cimento, **3**, 845, 1956.
91. Ovsyannikov, L. V., Doklady Akad. Nauk, **109**, 112, 1956.
92. Callan, C., Phys. Rev. D, **2**, 1542, 1970.
93. Symanzik, K., Comm. Math. Phys., **18**, 227, 1970.
94. Hooft't G., Report at the Conference on Lagrangian Field Theories, Marseille, 1972.
95. Gross, D., Wilczek, F. Phys. Rev. D, **8**, 3633, 1973.
96. Politzer, H.D., Phys. Reports, **C14**, 129, 1974.
97. Polyakov, A. M., Phys. Lett., **B59**, 82, 1975.
98. Wilson, K., Erice Lectures. CNLS-321, 1975.

NOTATION

(x, y) are points in the Minkowski space, and $(\boldsymbol{x}, t)$, $(\boldsymbol{y}, s)$ or $(\boldsymbol{x}, x_0)$, $(\boldsymbol{y}, y_0)$ are their space and time components, respectively.

The metric tensor $g^{\mu\nu}$ has the form diag$(1, -1, -1, -1)$.

All vectors are assumed to be covariant; $ab = a_\mu b_\mu = a_0 b_0 - (\boldsymbol{a}, \boldsymbol{b})$ is the scalar product of the four-dimensional vectors a, b with the components $a_\mu b_\mu$; the components of four-dimensional vectors are labeled by Greek letters, and of three-dimensional vectors by Latin letters.

The constants $\hbar$ and c, unless otherwise stated, are taken to be equal to unity.

INDEX

INDEX

QC793 224523
.3
.F5 Slavnov
.S5213 Gauge fields, introduction to quantum theory